Home How-To
Handbook
Trim

RICK PETERS

STERLING

New York / London
www.sterlingpublishing.com

STERLING and the distinctive Sterling logo are registered trademarks of Sterling Publishing Co., Inc.

Library of Congress Cataloging-in-Publication Data available

10 9 8 7 6 5 4 3 2

Published by Sterling Publishing Co., Inc.
387 Park Avenue South, New York, NY 10016

© 2008 by Sterling Publishing Co., Inc.

Distributed in Canada by Sterling Publishing
c/o Canadian Manda Group, 165 Dufferin Street,
Toronto, Ontario, Canada M6K 3H6
Distributed in the United Kingdom by GMC Distribution Services,
Castle Place, 166 High Street, Lewes, East Sussex, England BN7 1XU
Distributed in Australia by Capricorn Link (Australia) Pty. Ltd.
P.O. Box 704, Windsor, NSW 2756, Australia

Sterling ISBN-13: 978-1-4027-4811-0
        ISBN-10: 1-4027-4811-6

Book Design: Richard Oriolo
Photography: Christopher J. Vendetta
Page Layout: Sandy Freeman
Illustrations: Bob Crimi
Contributing Editor: Cheryl Romano
Copy Editor: Barbara McIntosh Webb
Indexer: Nan Badgett

For information about custom editions, special sales, premium and corporate purchases, please contact Sterling Special Sales Department at 800-805-5489 or specialsales@sterlingpub.com

Contents

Introduction

TRIM IS SO MUCH MORE THAN merely "woodwork in a building, especially around openings," as Webster's says. When you think about it, it's the interior and exterior trim of a home that often defines its style. What would a Victorian home be without its turned columns and gingerbread trim? Or a Craftsman bungalow without its simple window casings and exposed joinery?

It's the trim that gives a home personality. Want a fresh look for a room? Change the trim. Bored with your home's exterior? Dress it up with trim. Replacing and installing trim is much easier today than in the past, thanks to new materials like urethane foam, and pre-made parts. If you can cut a piece of wood and hammer a nail, this book will show you how you can transform a room or even your entire home with trim.

How to Use This Book

There are roughly three sections: basics, projects, and troubleshooting. The first section (chapters 1–4) starts with *Trim Basics*. We cover the benefits of trim and then take you through the many trim types, styles, and typical installations. In *Trim Materials* we guide you through the vast array of materials, parts, and products on the market. *Tools* explores the everyday and specialty tools you'll need to work with trim. In *Trim Know-How* we'll teach you the skills you'll need to install trim with confidence: measuring and layout, cutting, coping, attaching, and finishing trim.

The second section, on projects (chapters 5–8), works from the ground up, beginning with *Floor Trim:* how to install various types of baseboard. Then there's *Wall Trim,* including chair rail, picture/plate rail, wall frames, wainscoting, and pilasters and columns. In *Ceiling Trim* you'll find ways to dress up your ceilings with molding and ceiling medallions. *Windows and Doors* opens up details for not only trimming windows and doors but also removing and installing them. The final section is *Troubleshooting and Repair,* where we show you how to diagnose and solve common problems involving trim.

Codes and Permits

If any of your projects involve adding, extending, or modifying electrical or plumbing lines as well as framing (particularly on exterior walls), check with your local building inspector for permit and inspection requirements. Usually, an inspector will first check your work at the "rough-in" stage (no wall coverings in place) and again when all the finish work is done. By making sure your work is done to code, an inspector helps protect both your family and your home.

1

Trim
Basics

S TRIM A FINISHING TOUCH to improve a home's
looks? Or is it a handy way to cover up a flaw or keep
out drafts? The answer is "Yes" to both questions.
Trim or molding is more than just applied ornamentation that dresses up a home's exterior or interior spaces.
In many cases, the real purpose of trim is to conceal
gaps and/or create airtight seals. In this chapter, we'll
look at the benefits of trim, types and styles, and how
trim is applied.

The Benefits of Trim

Trim can be practical, decorative, or both. Most trim serves both as ornamentation and to hide gaps or create seals.

HIDE GAPS. Most modern homes are stick-built; that is, a framework of 2-by material serves as a skeleton for the house. This skeleton is then covered with interior and exterior sheathing to create a rigid structure. Wherever there's a transition from sheathing to another material like flooring, or window and door openings, there will be a gap between the two materials, as illustrated in the top drawing. Various types of trim can bridge the gap and conceal it.

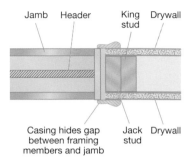

Jamb   Header   King   Drywall
                        stud

Casing hides gap   Jack   Drywall
between framing   stud
members and jamb

**TOP VIEW OF TYPICAL
DOOR OR WINDOW**

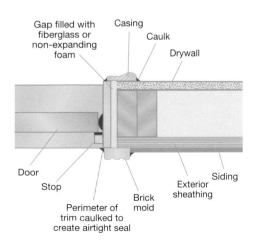

Gap filled with   Casing
fiberglass or   Caulk
non-expanding
foam   Drywall

Door
  Stop   Siding
                     Exterior
             Brick   sheathing
Perimeter of mold
trim caulked to
create airtight seal

CREATE AIRTIGHT SEALS. Gaps between sheathing and windows and doors on exterior walls are usually filled with insulation. In days past, fiberglass insulation was packed in the gap; today, most gaps are filled with expanding foam (see page 34). To further help prevent drafts, the edges of the trim are sealed with caulk once they're installed, as illustrated in the bottom drawing.

CREATE VISUAL INTEREST. Whether practical or not, trim can add a tremendous amount of visual interest to any room. Notice how the coffered ceiling and the window casing add to the richness of the living space in the photo below. Likewise, the large crown molding and picture rail add charm to the parlor shown in the top photo. Without trim, rooms tend to look flat and ordinary.

PRACTICAL AND DECORATIVE. In addition to door and window casings that conceal gaps and create seals, there are many other trim types—especially chair rail and wainscoting—that are both practical and decorative.

CHAIR RAIL. The practical side of chair rail is that it prevents damage to a wall when a chair bumps up against it. You'll find chair rail most often in dining rooms and kitchens. For aesthetic purposes it's typically run around the entire perimeter of the room. In a decorative sense, the chair rail also helps to break up large walls while providing an accent.

WAINSCOTING. Similar to chair rail, wainscoting protects against chairs, but it also provides protection all the way to the floor. If left natural, the richness of wood can add warmth and depth to a room.

Typical Trimming Sequence

Besides window and door trim (see Chapter 8), the most common trim installed is perimeter trim. This wraps around the perimeter of a room; think of baseboard, chair rail, or crown molding. The sequence for installing perimeter trim is pretty much the same. You start on the most visible wall and cut the trim to fit between the walls (these are butt joints, described on page 20). Moving on to a side wall, you'll cope one end to fit against the first piece and butt the other end against the opposite wall, as illustrated in the drawing below. Coping or making a coped joint entails shaping the end of trim to fit against the profile of previously installed trim (see page 21 and pages 80–82 for more on coping). Work continues by butting and coping trim all the way around the perimeter.

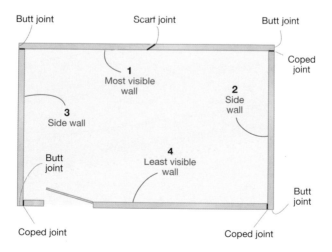

TRIM SEQUENCE

Types of Trim

If you've ever wandered down the trim or molding aisle at a home center or lumberyard, you know that there's almost too much choice in molding profiles, shapes, sizes, and materials. (We'll cover material choices in Chapter 2.) Trim is most often classified by its function, such as casing, baseboard, or crown. Regardless of its function, trim can be a single piece or multiple pieces (called a buildup by the pros). And, the trim can be installed either flat or at an angle.

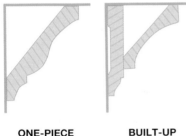

ONE-PIECE MOLDING   **BUILT-UP MOLDING**

ONE-PIECE VERSUS BUILT-UP. Modern trim manufacturers can produce large and richly sculpted one-piece trim. In the past, complex profiles were usually formed by attaching smaller trim pieces together. Buildups are still popular today, because they often make it easier to install the trim (see page 90).

FLAT VERSUS ANGLED. Trim is installed either flat against a surface or at an angle (often called a "sprung" molding by pros). Flat trim is simple to install, while angled trim is more of a challenge because, naturally, you have to cut and install it at an angle (see pages 76–77 and 158–163, respectively, for more on cutting and installing angled trim).

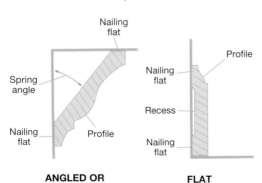

ANGLED OR "SPRUNG" MOLDING   **FLAT MOLDING**

CASING. Any trim that surrounds a door or window is called casing. It can be flat or highly profiled. Window casing is often installed with mitered corners (as shown here and described on page 20). Alternatively, if the window has a stool, the side trim butts up against the stool and is mitered on top (as illustrated

on page 170). Door casing can be mitered on top or can butt up against rosettes or plinth blocks, as illustrated in the drawing on page 176.

CEILING TRIM. Several types of trim fall under the ceiling trim umbrella: Crown molding, ceiling medallions, and coffered ceilings are just a few. All of these draw the eye upward and create visual interest. A ceiling medallion can also serve to conceal ceiling damage or gaps resulting from differences in fixture mounting requirements.

CORNICE. A cornice, broadly speaking, is a molding or group of moldings at a corner. In most cases the corner is at the junction of wall to ceiling, like the ornate cornice shown here. In bygone times, complex moldings like this were actually hand-made of plaster and were quite expensive. The cornice shown here is made of high-quality urethane foam by Fypon (www.fypon.com) and is a snap to install.

COLUMNS. Columns (also called pillars) can be decorative or structural, round or square, tapered or straight. Decorative columns are made of wood or foam. Structural columns are made of solid wood or metal. Plain metal column supports can be wrapped with a decorative column, as described on pages 148–151.

PILASTERS. A pilaster is a shallow column that's attached to a wall or cabinet to create the illusion that it's providing support. Pilasters are only decorative and so can be made of wood or foam.

WAINSCOTING.

Whenever trim covers the lower portion of a wall, it's generally called wainscoting. The most common forms of wainscoting include tongue-and-groove (shown on pages 126–132), quick-install wainscoting, (shown on pages 133–135), and flat-panel

wainscoting (as described on pages 136–141). Wainscoting can be left natural or be stained or painted.

FIREPLACE SURROUNDS.

Many trim manufacturers make trim "systems" for creating a fireplace surround. These include plinth blocks, rosettes, vertical and horizontal trim, and mantels.

Most of these systems require woodworking skills. If you like the look but don't want to hassle with building a surround, you can buy one complete—the surround shown in the bottom right photo is a one-piece unit manufactured by Fypon (www.fypon.com).

Trim Styles

Different trim profiles can be selected and grouped together to mimic a style such as Colonial, Federal, Victorian, Arts & Crafts, or modern. What's interesting is that many of these distinctive styles use some of the same trim. How the trim is used sets the styles apart.

COLONIAL. Original Colonial interiors were quite austere. Walls were typically plaster, and framing was often left exposed. Trim, if there was any, was simple. Modern Colonial interiors often utilize simple trim such as flat window casing and simple wall frames.

FEDERAL. Interiors trimmed in the Federal style tend to be light and airy. They usually include high ceilings and feature bold cornices and large windows. Stout columns and pilasters are also a favorite, like the pilasters defining the door opening shown here.

VICTORIAN. The original Victorian interiors tended to be over the top— extremely ornate, highly detailed trim featuring spires, towers, spindles, and brackets. Modern Victorian interiors are more subdued, although still often festooned with complex moldings and brackets.

ARTS & CRAFTS. The trim used for Arts & Crafts or Craftsman-like interiors is virtually the opposite of Victorian trim. Lines are clean and simple; curves, if any, are long and graceful. The trim is almost always left natural to highlight the warmth and beauty of the wood.

MODERN. "Eclectic" best describes trim used for a modern interior. Like Arts & Crafts, modern trim tends toward the clean and simple.

How Trim Is Attached

For the most part, trim attaches to underlying framing with nails. Whenever possible, it is also attached to adjacent wood surfaces such as the jambs of a window or door.

WINDOW TRIM. The casing for a window is usually held in place with nails of two different lengths. Longer 2" to 2½" casing nails are driven through the outside edges or perimeter of the casing, into the jack stud under the sheathing, as illustrated in the top drawing. Shorter 1½" finish nails secure the inner edges to the jambs, as shown.

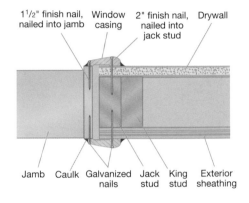

1½" finish nail, nailed into jamb   Window casing   2" finish nail, nailed into jack stud   Drywall

Jamb   Caulk   Galvanized nails   Jack stud   King stud   Exterior sheathing

DOOR TRIM. Door casing is attached much like window casing. The only difference is that longer nails are used to attach the thicker brick mold on the exterior, as illustrated in the bottom drawing. Note that for both window and door exteriors, hot-dipped galvanized nails should always be used.

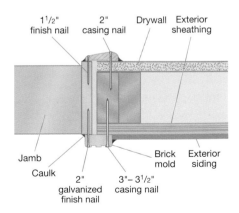

1½" finish nail   2" casing nail   Drywall   Exterior sheathing

Jamb   Caulk   2" galvanized finish nail   3"–3½" casing nail   Brick mold   Exterior siding

WALL AND FLOOR TRIM. Attaching wall trim can be a challenge since wall studs are spaced 16" apart on center. This means the trim can only be attached where studs are located. Floor trim is much easier to attach, as it has an additional nailing surface—the sill or sole plate.

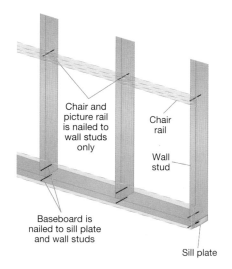

Chair and picture rail is nailed to wall studs only

Chair rail

Wall stud

Baseboard is nailed to sill plate and wall studs

Sill plate

TRIM BASICS

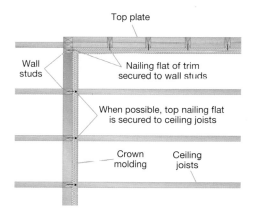

Top plate

Wall studs

Nailing flat of trim secured to wall studs

When possible, top nailing flat is secured to ceiling joists

Crown molding

Ceiling joists

CEILING TRIM. Ceiling trim is the most complex to attach because ceiling joists run in only one direction. On walls that are perpendicular to the joists, you can pin the top of the trim to the ceiling joists, as shown. Walls that run parallel to the joists do not allow for this option, so another way to secure the trim is required (see pages 90–91).

Basic Trim Joints

The most common ways to join together pieces of trim are butt, miter, compound miter, scarf, coped, and biscuit joints, as illustrated in the drawings on this page and the opposite page.

BUTT JOINT. Although the butt joint is the weakest of wood joints, it is common in trim work. A butt joint is made by simply butting two pieces of wood together. Without fasteners, this joint offers virtually no mechanical strength. But with fasteners, such as nails or screws, a butt joint provides adequate strength—and it's quick.

BUTT JOINT

MITER JOINT. A miter joint is made by making angled cuts on the ends of connecting pieces so that no end grain shows. Miter joints are one of the most common ways to join together trim, such as baseboard, casing, and crown molding.

MITER JOINT

COMPOUND MITER JOINT. On a power miter saw, a compound miter joint is made by both pivoting and angling the blade to create a cut that's angled in two directions. This is common practice when installing a "sprung" molding (as described on page 12), such as crown molding.

**COMPOUND
MITER JOINT**

SCARF JOINT. Whenever trim needs to be spliced together to create longer lengths, a scarf joint is used. The ends of the connecting pieces are mitered at opposing angles to mesh with one another. This creates a nearly invisible seam.

SCARF
JOINT

COPED JOINT. Coped joints are used extensively by savvy trim carpenters. There are two halves to a coped joint. On one half, the molding profile is left intact and simply butted into the corner of a wall or cabinet. The second half is the part that's coped to fit the profile of the molding butted into the corner. When cut properly, a coped part will butt cleanly up against the molding profile with no gaps. Not only will this joint fit nicely when completed, but it will also stay that way: There's less exposed end grain than with a miter joint, which will expand and contract as the seasons change. Miter joints are notorious for opening and closing as the humidity changes, producing a gap of varying widths throughout the year.

COPED
JOINT

BISCUIT JOINT. A biscuit joint is similar to a butt joint except that it is strengthened by the addition of a biscuit. The biscuit—a football-shaped piece of compressed wood—fits into half-moon-shaped slots dished out of both ends of the adjoining pieces. These slots are formed with a special tool called a biscuit jointer. When glue is applied to the biscuit, it swells and locks the pieces together. (For more on biscuit joinery, see pages 88–89.)

BISCUIT
JOINT

2

Trim Materials

THE TERM "TRIM MATERIALS" might seem need-less—after all, trim is made from wood, right? What's to know? The fact is that less than half of the trim at most home centers and lumberyards is actually made of wood. New materials—especially urethane foam—are gradually taking over the trim market. These substances don't warp and twist as wood does, plus they're more flexible and weather-resistant. Let's look at all these materials in detail.

Softwood Lumber

Wood—either softwood or hardwood—is still used to make much of the trim on the market. In some cases you'll need hardwood or softwood lumber for a job such as installing a foundation for crown molding or installing wainscoting. To purchase lumber wisely, you should know about the different grades of softwood (discussed here) and hardwood (opposite page) lumber.

Since most softwood lumber sold is used in the construction industry, softwood grading takes into account strength, stiffness, and other mechanical properties. The problem is that no two woods have identical characteristics. This means that every softwood species has its own set of grading guidelines. The most common grades of pine are listed below. For trim jobs where the wood will be painted, go with D select; for clear topcoats, choose either B&BTR or Superior grade.

Softwood Grades

CLASSIFICATION	GRADE	APPEARANCE
Select	B&BTR	Many pieces are absolutely clear and free from knots; only minor defects and small blemishes are permitted.
	C select	Small defects and blemishes allowed. Recommended for all finishing uses where fine appearance is essential.
	D select	Defects and blemishes are more pronounced; used when finishing needs are less exacting.
Finish	Superior	Only minor defects and blemishes allowed.
	Prime	Similar to Superior but with more defects and blemishes allowed.
	E	Pieces where crosscutting or ripping will produce Superior or Prime grades.
Common	#1 common	The ultimate in fine appearance in a knotty material; all knots must be small and sound.
	#2 common	Contains larger, coarser defects and blemishes; often used for knotty-pine paneling.

Hardwood Lumber

If your trim job calls for hardwood lumber, you'll need to know how it's graded. The common grades that you'll encounter are FAS (Firsts and Seconds), select, No. 1 common, and No. 2 common (actually 2A and 3A; see the chart below). Basically, the grade of a piece of hardwood depends on how much clear wood the board will yield in relation to its total square footage or surface measure. Each grade specifies this as a percentage: roughly 83% for FAS and Select, around 67% for No. 1 common, and 50% for No. 2 common. One important thing to realize about hardwood grading is that the clear wood in a No. 2 common board is the same *quality* as that in an FAS board—there's just less of it.

Hardwood Grades

	FAS	SELECT	#1 COMMON	#2A & 3A COMMON
Minimum size board	6" × 8'	4" × 6'	3" × 4'	3" × 4'
Minimum size cutting	4" × 5' 3" × 7"	good face grades FAS, poor face grade #1 common	4" × 2' 3" × 3'	3" × 2'
Basic yield	83%	83%	67%	50%

Sheet Goods

Some trim jobs— primarily wain- scoting—will call for sheet goods. Common sheet goods, shown in the photo above, are hardboard, plywood, and MDF (medium-density fiberboard). Grade in softwood plywood refers to the quality of the veneer used for the face and back veneers (A-B, B-C, etc.); see the chart below. Stick with AC plywood, and place the A face out.

The different grades of hardwood plywood describe only its appearance, not its core type or strength. Choose hardwood plywood only if you're using a clear topcoat, then pick a species to match your trim—oak and birch plywood are commonly available.

MDF (medium-density fiberboard) is an engineered product. Wood fibers are coated with resin and compressed to form sheets. Since there's no grain, changes in humidity have little effect on MDF. And this means stability. The small particles also create a solid, homogeneous edge that takes paint well and machines easily.

Softwood Plywood Grades

VENEER GRADE	CHARACTERISTICS
A	Smooth, paintable. Not more than 18 neatly made repairs permitted: boat, sled, or router type, and parallel to grain. Wood or synthetic repairs permitted. May be used for natural finish in less demanding applications.
B	Solid surface. Shims, sled or router repairs, and tight knots to 1" across grain permitted. Wood or synthetic repairs permitted. Some minor splits permitted.
C	Tight knots to 1½". Knotholes to 1" across grain and some to ½" if total width of knots and knotholes is within specified limits. Synthetic or wood repairs. Discoloration and sanding defects that do not impair strength permitted. Limited splits allowed.
D	Knots and knotholes to 2½" width across grain and ½" larger within specified limits. Limited splits allowed. Limited to Interior, Exposure 1, and Exposure 2 panels.

Trim

Trim can be made from softwood, hardwood, foam, plastic, or medium-density fiberboard. Softwood trim is further broken down into two grades: paint-grade and stain-grade; see below.

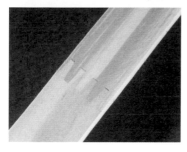

PAINT-GRADE. Most softwood trim is not machined from a single piece of softwood. Instead, it's formed by splicing together smaller pieces, as shown in the top photo. This does two things: It makes the trim cheaper to manufacture, and it also helps reduce warp and twist caused by variations in grain. To make paint-grade trim, manufacturers machine finger joints into pieces to be joined, as shown in the middle photo. When glued, this creates a surprisingly strong joint.

STAIN-GRADE. Trim that is stain-grade is made from a solid piece of wood and can be stained or finished with a topcoat, as shown in the bottom photo. If you were to apply a stain or topcoat to a piece of paint-grade trim, the variations in wood and also the joints would be quite evident. Because it's made from a single piece of wood, stain-grade trim is more expensive than paint-grade. Buy it only when you plan on applying a stain or clear topcoat.

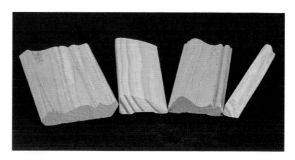

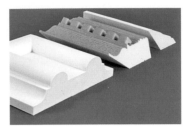

FOAM. When foam molding first hit the home centers, it wasn't very well received. The problem was the low density of the foam. Advances in materials and production technology now allow for better foam moldings. In fact, today's foam offers many advantages over wood. It machines like butter, it's flexible, it's stable (no warp or twist whatsoever), and it's basically impervious to weather. If you love the look of natural wood trim, look into foam. Savvy trim manufacturers like Fypon (www.fypon.com) now offer stainable foam molding that is textured like wood (see pages 158–163 for an example of this). Once it's stained, you'd swear it's natural wood.

PLASTIC. Plastic molding is most often made from PVC (polyvinyl chloride). Like its plumbing cousin (PVC pipe), PVC trim is impervious to water and so is great for exteriors or damp locations. Its only drawback is color: It comes in white and is tough to finish in other colors.

QUICK FIX

Pre-Made Corners

If you like the idea of installing trim but are leery about cutting miters to join pieces at the corners, consider using pre-made corners. The ones shown here are for crown molding (one of the more challenging types of trim to install) and are made by Fypon (www.fypon.com). Just secure one to an inside or outside corner and then butt the crown molding up against it—no joints to cut.

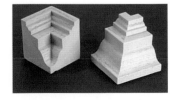

MDF. Another trim material that's gaining popularity is MDF (medium-density fiberboard), as shown in the top photo. Just like the sheet stock (page 26), trim made of MDF is stable, takes paint well, and won't warp or twist. It's usually sold already primed and ready for paint.

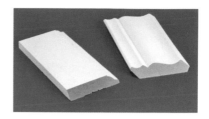

LAMINATED TRIM. In the home center molding aisle, you'll also find a variety of trim that's laminated with a thin plastic, paper, or foil covering designed to imitate another material, as shown in the middle photo. Wood-grained coverings are the most popular but unlike real wood cannot be stained; so make sure you choose the exact color you're after.

PRO TIP

Pre-Finished Trim

Some lumberyards and home centers sell pre-finished trim. Not only does this eliminate the mess of finishing the trim, but it also offers durability: A manufacturer-applied finish is often baked on and is much hardier than a hand-applied finish. If you can't find pre-finished trim at your local home center, check with the folks in the kitchen and bath design center—virtually all cabinet manufacturers make a wide variety of pre-finished trim that can be ordered and shipped directly to you.

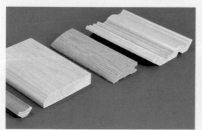

Trim Accessories

In addition to lengths of trim, most home centers and lumberyards sell trim accessories like corner blocks, rosettes, and other pre-formed parts.

CORNER BLOCKS. Corner blocks take the hassle out of going around corners. They are shaped to handle both inside and outside corners. In use, you simply install one at a corner and then butt the trim up against it.

ROSETTES. A rosette is a square block of wood with a design in its center and is used to handle vertical-to-horizontal trim transitions. An example of this is wrapping trim around a door opening, as described on pages 176–179.

PRE-FORMED PARTS. In addition to corner blocks and rosettes, you'll likely find other pre-formed trim parts in the trim aisle. The most common of these is the plinth block (far right in bottom photo). A plinth

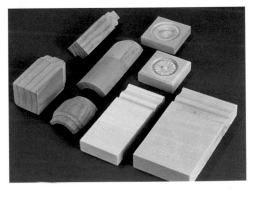

block is a rectangular block that serves as the base for a vertical trim piece such as door casing or a pilaster.

Fasteners

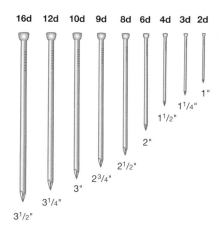

16d 12d 10d 9d 8d 6d 4d 3d 2d

Most trim is fastened in place with nails. Nails are sized using the antiquated "penny" system (abbreviated as "d"). This system is based on how much 100 nails used to cost (talk about inflation!). Penny is now used to indicate the length of the nail; see the drawing at right. Nails can be bought in 1-, 5-, and 50-pound boxes or in any quantity from a retailer that stores their nails in open bins. Exterior nails should always be hot-dipped galvanized to prevent corrosion and staining.

1"
1¹/₄"
1¹/₂"
2"
2¹/₂"
2³/₄"
3"
3¹/₄"
3¹/₂"

PRO TIP

Air-Driven Fasteners

Professional trim carpenters generally use air nailers to fasten trim in place (see pages 53–54) because they both drive and set the fastener in the blink of an eye. Air nailers use special fasteners that come in sticks. These sticks can be straight or angled, as shown here.

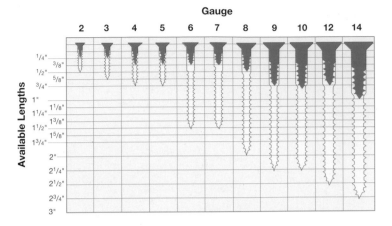

	Gauge										
Available Lengths	2	3	4	5	6	7	8	9	10	12	14
1/4"											
3/8"											
1/2"											
5/8"											
3/4"											
1"											
1 1/8"											
1 1/4"											
1 3/8"											
1 1/2"											
1 5/8"											
1 3/4"											
2"											
2 1/4"											
2 1/2"											
2 3/4"											
3"											

A screw is your best choice when strength is needed. Screws can be used to fasten foundation moldings to wall studs, or backer strips to wall studs (see page 129), and to mount load-bearing trim (like the plate rail described on pages 121–123). Regardless of the screw, whenever possible select square drive heads versus slotted or Phillips-head screws. The unique square recess in the screw heads (and matching screwdriver or driver bit) greatly reduces the tendency of a standard bit to "spin" while driving in a screw.

P R O   T I P

Trim-Head Screws

There's a special breed of screw designed for attaching trim. On a trim-head screw, the head is roughly one-half the size of a standard screw head. This little head allows you to substitute a screw in place of a nail whenever you need extra holding power.

Sealants

Caulk, window wrap, and expanding foam are all used to seal openings around windows and doors. Caulk is also commonly used to conceal gaps between trim and adjacent walls.

CAULK. There are many types of caulk available for sealing trim; see the chart below. The two most common are acrylic latex and silicone, as shown in the top photo. Acrylic latex can be painted and is used to fill gaps between trim and interior walls. Silicone caulk seals gaps on exterior walls. Make sure to always use 100% silicone caulk for maximum flexibility.

Caulk Types/Applications

MATERIAL	APPLICATION
Acrylic latex	Most common application is filling gaps and voids behind trim; also know as painter's caulk, as it accepts paint well; inexpensive and very easy to use and clean up.
Butyl	Applications where you need to seal metal (such as flashing around a window or door); inexpensive, flexible, but very messy to use (wear disposable rubber gloves).
Silicone	Exterior applications that won't be painted; extremely flexible and long-lasting, but can't be painted.
Siliconized acrylic latex	Interior and exterior applications where a more flexible caulk is needed; bonds better to surfaces and lasts longer than acrylic latex; less messy and easier to use than silicone.

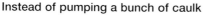

QUICK FIX

Caulk Saver Rod

Occasionally you'll find a large gap that needs filling. Instead of pumping a bunch of caulk into the gap, consider first installing some foam rod (often called caulk saver), as shown here.

WINDOW AND DOOR WRAPS. Any opening in an exterior wall is a potential source of leaks. If your home isn't sealed with a whole-house wrap (such as Tyvek), consider applying window or door wrap around an opening before installing a new door or window. Self-adhesive moisture wraps are available wherever windows and doors are sold. These are applied to the framed opening and overlap over the sheathing (under the siding) to create a watertight seal. (See page 170 for more on this.)

Window-and-Door Foam

★ In modern construction, spray-on foam is the number one way to fill gaps. It's important that you choose the correct type, however, when insulating around windows and doors, as described below.

STANDARD FOAM. Standard expanding foam will expand to roughly twice its original size. It's best used to fill gaps around framing members, and it dries rigid.

WINDOW-AND-DOOR FOAM. Window-and-door foam is specially formulated to expand less than standard foam— and it's more flexible when dry. While standard foam tends to bow window and door jambs, window-and-door foam does not.

Adhesives

Although most trim is held in place solely with nails, you'll often find it wise to further secure the trim with an adhesive. The three most widely used adhesives for this are carpenter's glue, construction adhesive, and silicone caulk. Carpenter's or yellow glue is a cross-linked PVA (polyvinyl acetate). A common brand of this is Titebond. It sets up quickly (about 15 minutes), and clamps need to be applied for only around 1 hour. Construction adhesive fills gaps, sets up without clamps, and is surprisingly strong. Although most folks don't consider silicone an adhesive, it offers excellent adhesion properties, combined with virtually permanent flexibility. Both silicone caulk and construction adhesive come in cartridges that fit into standard caulk guns.

PRO TIP

Polyurethane Glue

Polyurethane glue is a newcomer that is extremely versatile. It works particularly well with foam and plastic trim. It does have its disadvantages, though. If it gets on your skin, there's not a solvent around that'll take it off—make sure to wear rubber gloves. And, although a bead of polyurethane glue will expand 2 to 3 times its size to fill gaps, the foamy substance has absolutely no strength.

Finishing Supplies

When it's time to apply a finish to trim, your choices include paint, stain, and clear topcoats. You'll also need to deal with filling nail holes. Which material you use for this depends on *when* you fill the holes—before or after the finish is applied, as described below.

PUTTY. Not all putty is the same. There are two main types: hardening (top left photo) and nonhardening (top right photo). Hardening putty is applied prior to finishing and is typically sanded smooth. Nonhardening is color-matched to the finish on your trim (usually natural or stained wood) and is used after the finish has been applied—it does not harden over time.

WAX CRAYONS. Many trim carpenters and most cabinet installers use wax crayons to fill nail holes. This is done after the finish is applied—as most cabinet trim is pre-finished. They're available in a wide variety of colors and shades and are easy to use (see page 97).

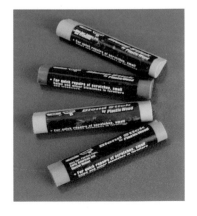

STAIN AND CLEAR TOPCOATS.

Wood can be left natural or be stained. Stains let you alter the appearance of the wood. You can either tint the wood with semi-transparent stains, which allow you to still see the wood's grain, or use an opaque stain for a complete coverup. Whether stained or left natural, the trim will also need to be sealed with a clear topcoat. Clear topcoats include polyurethanes and

spar and marine varnishes. Each of these penetrates into the wood and lets the natural beauty show through. Spar and marine varnishes have more "solids" to create a thicker, more durable coat.

PAINT. Paint is the way to go when you want to cover up trim (such as paint-grade trim) or if you want to add color to your trim. Always start with a coat of primer before applying the top coat of paint.

PRO TIP

Sanding Sealers

When dealing with woods that have a reputation for being finicky to finish (such as pine), pros often reach for a sanding sealer. These products seal wood grain prior to applying a topcoat and help reduce splotching. They can be sanded when dry and therefore should be used only on unstained wood trim.

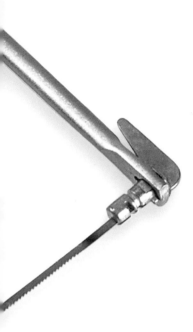

3

Tools

CHANCES ARE THAT YOU ALREADY OWN many of the tools needed for working with trim: A tape measure and level, a hammer and nail sets, and a saw are in most toolboxes. In this chapter we'll cover all the basic tools, plus some specialty items that can make installing trim a lot easier: a power miter saw, jigs for cutting crown molding, and air nailers.

Measuring and Marking

The measuring and marking tools you'll need to accurately lay out trim include: a tape measure, framing and combination square, compass, stud finder, chalk line, contour gauge, level, and angle gauge.

TAPE MEASURE. When you're buying a tape measure, go with a name brand you can trust. Since you'll often be extending the tape out over long distances when measuring for trim, consider one of the wider-blade tapes, like the FatMax series by Stanley.

FRAMING SQUARE. A framing square is useful for checking walls, floors, and ceilings for square, as well as for laying out trim. We prefer an aluminum square to a steel one since it's much lighter and won't rust over time.

COMBINATION SQUARE. A combination square is great for marking reveals (see page 64). It's basically a metal rule with a groove in it that accepts a pin in the head of the square. The head has two faces—one at 90 degrees, the other at 45 degrees. When the knurled nut on the end of the pin is tightened, the head locks the rule in place at the desired location.

COMPASS. A compass is useful for bisecting angles (as described on page 66) and for scribing parts (as described on page 82). The most common type is the wing compass that has legs hinged at the top. One leg has a steel point and the other holds a standard pencil.

ELECTRONIC STUD FINDER. You can't attach trim to a wall or ceiling unless you can locate its studs or joists. An electronic stud finder is the tool for the job. Recent advancements in technology have driven down the price of these tools while boosting their accuracy.

CHALK LINE. Chalk lines are useful for marking long, straight lines. Make sure to shake the chalk line before use to evenly coat the line. Hook one end at your mark, stretch the line to the opposite end, and pull the line straight up a couple of inches; then let go to "snap" a line.

CONTOUR GAUGE. A contour gauge lets you accurately copy the irregular shape of an obstacle such as a pipe or door casing and then transfer it onto your trim.

LEVELS. The most common levels are 3 or 4 feet long, with bodies made of wood, metal, or plastic; smaller torpedo levels are useful in tight spots. Virtually every level has multiple vials, which are curved glass or plastic tubes filled with alcohol (hence the name "spirit" level). A bubble of air trapped in the vial will always float to the highest point on the curve.

Laser Levels

Many trim jobs require drawing a line around the perimeter of a room for positioning the trim. You could snap a chalk line, but then you'd have to remove the chalk from your walls. A better method is to use a laser level. These are coming down in cost, and can be rented at most home and rental centers. The big advantage of a laser level is that it shoots a perfectly level line along a wall—or around the perimeter of a room—without leaving any marks.

ANGLE GAUGE.
Designed specifi-
cally for meas-
uring the angle
of two adjacent
surfaces, plastic
angle gauges like
the one shown

here take all the guesswork out of cutting
trim to fit on out-of-square surfaces.

SLIDING BEVEL.
A sliding bevel is
invaluable for verifying

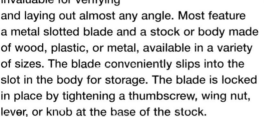

and laying out almost any angle. Most feature
a metal slotted blade and a stock or body made
of wood, plastic, or metal, available in a variety
of sizes. The blade conveniently slips into the
slot in the body for storage. The blade is locked
in place by tightening a thumbscrew, wing nut,
lever, or knob at the base of the stock.

PRO TIP

Digital Protractor

If you've ever installed trim in your home, you
already know that most walls aren't plumb and most
corners aren't square. The challenge to installing trim is to
identify the actual angles where walls and ceilings meet.
Bosch recently introduced the Miterfinder digital protractor.
This tool is actually four tools in one: an angle finder, a compound cut calculator, a protractor, and a level. (For more on
using this tool, see page 67.)

Hand Tools

Y ou'll need various hand tools to install trim, including a hammer, nail set, screwdriver, utility and putty knives, file, plane, chisel, and saw.

CLAW HAMMER. If you won't be using air power to drive in your fasteners (see pages 53–54), you'll need a quality hammer. Curved-claw hammers are by far the most common, and are especially good at removing nails. The most common weight is 16 ounces. A heavier hammer will drive nails faster, but will also make you tire faster.

NAIL SETS. The trick to nailing on trim effectively is knowing when to stop. The closer you get to the surface, the greater the chance of dinging it with a hammer. To eliminate the chance of dings, stop when the nail head nears the surface, then reach for a nail set. A nail set is designed to finish the job by driving the nail below the surface, or "setting" it.

SCREWDRIVERS. Multi-tip screwdrivers (like the top screwdriver in the photo at right) pack a lot of punch in a small space. At minimum, you'll want a 4-in-1 that features two different-sized Phillips and slotted bits. An offset screwdriver can get into places a standard screwdriver can't.

UTILITY AND PUTTY KNIFE. For many trim jobs, you'll find a utility knife handy. They're great for fine-tuning trim, cutting shims, and removing drywall.

PUTTY KNIVES. A putty knife is useful for removing trim—a pair of putty knives is one of the best ways to prevent damage to underlying walls when removing trim (as described on page 68). They're also used for removing old caulk and filling nail holes with putty.

FILES. Files come in many shapes and sizes. The most common shapes are: mill (or flat), half-round, round, 4-in-hand, and triangular. You'll find all of these shapes useful for fine-tuning trim, especially when making a coped profile (see page 81).

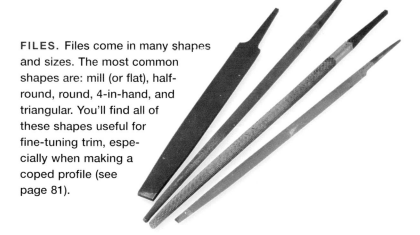

RIFLERS. Riflers are specialty files used primarily by carvers to smooth out small details in their work. These may be double-ended or come with a handle. Riflers are available individually or in sets and can be either files or rasps. They are particularly adept at getting into tight corners where ordinary files can't reach—especially when filing a coped joint.

BLOCK PLANE. A block plane is handy for fitting joints and trimming parts to fit, as described on page 83. Block planes are between 5" to 6½" in length; they were originally designed to plane ornery end grain (which they do extremely well).

HAND PLANE. A hand plane (commonly called a jack plane or jointer plane) is the perfect tool to fine-tune trim along its length. Many trim pieces need to be scribed to fit (page 83) so they'll fit snugly against an uneven wall or ceiling. A hand plane can quickly whittle away the unwanted wood. As a general rule, the longer the sole of a plane, the wider its blade.

CHISELS. Chisels are great for paring trim to make tight-fitting joints. A set of quality beveled-edge chisels like those shown here can handle all your paring needs.

TOOLBOX SAW. Toolbox saws (usually around 16" in length) are designed to fit in larger toolboxes. They have aggressive teeth, so they're useful mostly for rough cuts.

MITER BOX AND SAW. For the most part, miter saws have been replaced by the power miter saw (page 50). Miter saws are designed to fit in a miter box. If the blade is sharp and there's no play in the guides, they work quite well.

COPING SAW. Coping saws are used for cutting curves and coping molding (pages 80–81). The blade is held in a tensioned frame that is adjustable, and the blade can be pivoted to make it easier to follow a curve.

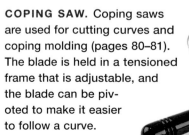

Power Tools

Although you can install trim with just hand tools, you'll find it a lot easier going with power tools. These include: electric drill and bits, biscuit jointer, saber saw, table saw, miter saw, router, and circular saw.

ELECTRIC DRILLS. The two types of electric drill you'll find useful for trim work are a cordless drill and a right-angle drill. Since trim work is often done in tight quarters, a right-angle drill lets you access those hard-to-reach places. Note how the drill chuck on a standard drill (right in photo) is in line with the motor, versus the chuck on a right-angle drill (left in photo), which is at 90 degrees to the motor.

DRILL BITS. Twist bits can drill into just about anything because they're ground to a "universal" angle, usually 118 or 135 degrees. When buying bits, look for high-speed steel and a reputable manufacturer. Stay away from bargain bits: They may be inexpensive, but they won't last.

BISCUIT JOINTER. A biscuit jointer (often called a plate jointer) cuts half-moon-shaped slots in wood to accept compressed wood biscuits. Biscuit joinery is quick and easy and is a great way to strengthen an otherwise weak butt joint. (See pages 88–89 for more on using this handy tool.)

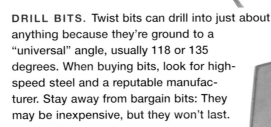

SABER SAW. Saber saws make quick work of cutting curves and, when used with the EasyCoper (see below), coping molding. There are two basic types of saber saws (commonly called jigsaws) on the market: standard and orbital. Quality saber saws will feature heavy-duty castings, variable speed, and orbital action. The orbital action creates an aggressive cut and should only be used for rough cuts.

TABLE SAW. A table saw is used for ripping, crosscutting, miters, bevels, and joinery. There are three main models of table saw to choose from: bench-tops saws, con-tractor's saws, and cabinet saws. For most trim work, a bench-top saw like the one shown here is all you'll need.

QUICK FIX

EasyCoper

Coping molding with a coping saw takes a steady hand, patience, and considerable skill. There must be an easier way, right? There is—It's called the EasyCoper. EasyCoper is a plastic aid designed by a professional car-penter that simplifies the coping of crown molding. The jig lets you use a saber saw instead of a coping saw to cope

molding. It holds both the molding and the saw at the perfect angle to make the cut. (For more on using this jig, see page 81.)

POWER MITER SAW. Power miter saws excel at mitering trim—especially crown molding—to exacting angles. Power miter saws come in three basic types: chop saw, compound, and sliding compound. On a chop saw, the blade pivots for angled cuts but doesn't tilt. The blade on a compound saw pivots and tilts, allowing for compound cuts. On a sliding compound saw, the blade slides back and forth on an arm to allow cuts on wider stock. (For more on using a power miter saw, see pages 71–77.)

MITER SAW STANDS. Although do-it-yourselfers often mount their miter saw to a workbench or other cabinet, that's not very convenient when it's time to use the saw on site. Most tool manufacturers make stands for miter saws. These range from simple fold-up stands to elaborate units with built-in fences and stops.

CROWN MOLDING JIGS. There are many crown molding jigs on the market that take the guesswork out of cutting crown. The crown jig shown here is made by Woodhaven (www.woodhaven.com); it holds the molding at its intended angle while you cut it. So, no more compound cuts and tedious trial and error. Woodhaven's jig requires only a simple 45-degree miter cut that any miter saw can handle.

ROUTER. If you want to make your own molding (see pages 94–95), you'll need a router. Routers fall into two categories: fixed-base or plunge. On a fixed-base router, the motor unit slides up and down or rotates within the base to adjust the depth of cut. With a plunge router, the motor unit slides up and down on a pair of spring-loaded metal rods. To make a cut, you release a lever and push down to lower (or plunge) the bit into the workpiece.

ROUTER BITS. There are two basic types of router bits: unpiloted (top middle photo) and piloted (bottom middle photo). Piloted bits are guided by a ball bearing mounted on the bit; unpiloted bits require some type of guide (like a fence or edge guide). Bits can be high-speed steel (HSS) or carbide-tipped. As a general rule of thumb, go with carbide. They cost more, but they'll stay sharper a whole lot longer.

PRO TIP

Router Table

Mounting a router in a table effectively turns it into a mini-shaper. You can accurately cut joints, work safely with small parts, and add a level of precision to your router work that may not have seemed possible.

CIRCULAR SAW. Circular saws are best used for rough-cutting trim to length and for cutting sheet stock to size. By far the most common homeowner saw is the shaft-drive, but the worm-drive type is favored in construction—it's rugged and easy to use. All circular saws have similar parts. The saw blade attaches to an arbor that is either the motor shaft or a gearbox (as with a worm-drive saw). A pivoting blade guard retracts up and out of the way during the cut and slips back down to cover the exposed blade once the cut is made. The base of the saw pivots from side to side for bevel cuts and from front to back to adjust the depth of cut.

PRO TIP

Cordless Trim Saw

A standard circular saw (above) is heavy and often cumbersome to use—especially when you're wrestling with a cord. A lighter, easier-to-use version is the cordless trim saw, like the one shown here. For starters, there's no cord to hassle with. Second, these saws weigh a lot less—typically around 6 pounds. They sport smaller blades ($3\frac{1}{2}$" to 5") compared to standard saws ($7\frac{1}{4}$") but can still cut through a 2×4 with ease. Although not as powerful as their corded cousins, they can handle most trim jobs.

Air Tools

If you're not handy with a hammer, you'll love an air nailer. Air nailers use compressed air supplied (in most cases) by a compressor (see below) to drive and set a fastener in the blink of an eye. Nailers commonly used to install trim include brad nailers, narrow-crown staplers, and finish nailers.

COMPRESSORS. There are two basic types of compressors: oil-lubricated and oil-less. Oil-lubricated compressors have a reputation for hardiness and dependability on the job site. Oil-free or "self-lubricated" compressors run without oil. Instead, they use non-metal piston rings, Teflon-coated parts, and sealed bearings. In effect, they're maintenance-free.

AIR HOSES. A standard air hose (black hose in photo) is made of PVC. It's relatively inexpensive and remains flexible over a wide range of temperatures. Clear plastic hose (white hose in photo) is popular in the trades because it's rugged yet lightweight. On the downside, it isn't as flexible as standard hose and has a habit of not lying flat; it can get tangled in your feet.

BRAD NAILERS. Brad nailers are perfect for securing lightweight and delicate trim such as shoe molding and cove molding. They typically shoot 18-gauge brads that vary in length from $5/8$" to $1\frac{1}{2}$".

NARROW-CROWN STAPLERS.
Narrow-crown staplers shoot ¼"
or ⅜" staples in gauges ranging
from 16 to 22 and that come in
varying lengths. Although they
offer better holding power than a
brad or nail, they have a larger
footprint. As it's difficult to hide
the heads of staples, narrow-
crown staples are used mostly in
areas that won't be seen.

FINISH NAILER. A finish nailer makes
any trim job a whole lot easier. A finish
nailer shoots 15- or 16-gauge nails that
vary in length from ¾" to 2¾". When
looking to buy a finish nailer, you'll first
need to decide on what gauge nail to
shoot. Thinner 16-gauge nails are less
likely to split wood than the heavier 15-
gauge. But the disadvantage to the
smaller gauge is that the nails tend to
follow the grain in wood, and often
deflect off course.

PRO TIP

Cordless Nailers

The power and precision of an air nailer—without a
compressor and hose—brilliant! That's what the folks
at Paslode (www.paslode.com) have been offering for years
(they pioneered cordless air nailers in 1986). A cordless nailer

uses a rechargeable battery and a
replaceable fuel cell to run a com-
bustion motor to drive fasteners.
Cordless nailers are a joy to use.
Once you've used one, you'll dread
going back to a standard nailer.

Safety Gear

With any project around the home, it's important to don appropriate safety gear. For trim work, this includes protecting your eyes, ears, lungs, and hands.

EYE AND EAR PROTECTION. Whenever you cut and install trim, you should always wear safety glasses. Chucks and slivers of wood

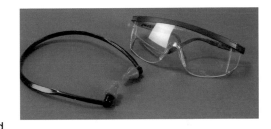

often fracture and fly off in all directions. Metal fasteners can also deflect and cause injury. The last place you want any of these is in your eyes. Also, if you're using power tools (including air nailers), protect your ears from the high-pitched whine of universal motors and the bang of compressed air.

RESPIRATOR. Sawdust is a known carcinogen. Protect your lungs from dangerous dust by wearing a respirator. If you're applying a finish to trim, it's best to wear a cartridge-style respirator to keep harmful vapors out of your lungs.

LEATHER GLOVES. Whenever you cut and handle wood trim, there's a possibility for slivers. So protect your hands with a pair of good-quality leather gloves.

Demolition Tools

Trim jobs often require some type of demolition. It can be as simple as removing old trim or as complex as knocking down a wall. Demolition tools include: prybar, drywall saw, cat's paw, end nippers, reciprocating saw, and locking-jaw pliers.

PRYBAR. A prybar is a short length of flat steel that's curved on its ends. The ends are also notched to make it easy to pull nails. Prybars are useful both for both prying off old trim and for pulling nails. A prybar tends to be less destructive to wall coverings than its beefy cousin the crowbar.

DRYWALL SAW. A drywall saw is the tool of choice for cutting and removing drywall. Its pointed tip makes it easy to drive the blade directly into the drywall, and the coarse teeth will chew right through the drywall as you saw.

CAT'S PAW. A cat's paw is a special prybar designed for removing nails flush with or below the surface of a workpiece. A hammer drives the sharp tip of the claw into the wood surrounding the nail. Then the tool is leveraged to pull out the fastener. Be aware that a cat's paw is a destructive tool, as it cuts into the surface to gain a purchase on the nail.

END NIPPERS. Originally designed for cutting the ends of wire flush, an end nipper is also an excellent tool for removing stubborn nails and brads. Simply grip the fastener with the sharp jaws of the nipper, and pivot to one side to lever out the fastener.

RECIPROCATING SAW. A reciprocating saw (commonly referred to by the brand name Sawzall) can make quick work of cutting through wall coverings and framing. When fitted with a demolition blade, it's the perfect tool for freeing windows and doors from rough openings prior to their removal.

LOCKING-JAW PLIERS. Better known by their brand name Vise-Grips, locking-jaw pliers are invaluable for removing fasteners. The opening of the serrated jaws is controlled by turning a knurled knob on the end of the wrench. What's special about this wrench is that the jaws will remain closed or "locked" around a fastener until a release lever is operated. Often this is the only tool that will let you pull out a really stubborn fastener.

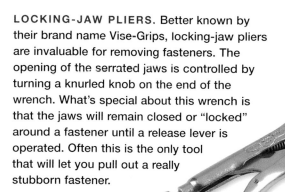

Trim
Know-How

THERE'S A LOT MORE to working with trim than
cutting and nailing, for one simple reason: You'll
rarely find two surfaces in a home that are level
and plumb. To achieve professional-looking results,
you'll have to take this into account as you work. In
this chapter, we'll look at the know-how required:
everything from measuring and layout tips, to working
with a power miter saw and attaching trim with air tools,
to making your own molding.

Checking Tools for Accuracy

Accuracy in measuring and layout is the foundation of any successful trim job. That's why it's important to periodically check your tools for accuracy—especially before starting a large job.

FRAMING SQUARE. To see whether a square is square, place one leg up against a known flat edge. Then draw a line along the blade and flip the square over as shown. Align the blade with the line you just drew—it

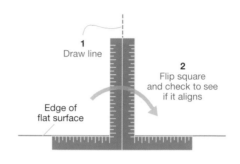

1 Draw line

Edge of flat surface

2 Flip square and check to see if it aligns

should be in perfect alignment. If it isn't, you can re-true the square by lightly hammering the metal at the diagonal joint between the tongue and the blade. If you hammer the inside edge, the square will "open"; hammering on the outside corner "closes" the square.

TAPE MEASURE. To check a tape measure, extend the tape out several feet and bend it back on itself. Align the inch marks and check to see whether the graduations are even. On cheaper tapes, you'll find that the graduations don't match up.

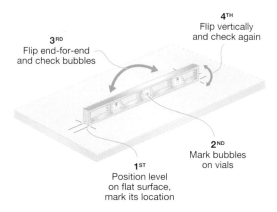

4ᵀᴴ
Flip vertically
and check again

3ᴿᴰ
Flip end-for-end
and check bubbles

2ᴺᴰ
Mark bubbles
on vials

1ˢᵀ
Position level
on flat surface,
mark its location

LEVEL. To check a level, place it on a known flat surface and mark its exact location with a pencil. Note the bubble reading on the vials with a pencil mark as well, and then flip the level end-for-end, aligning it with the pencil marks. Check the bubbles and then repeat by flipping the level upside down. All of the bubble readings should match. If they don't, most levels have adjustable vials to make this correction.

RULES. Check the graduations of your rules against each other—you may be surprised to find that they're not the same. The two rules in the bottom photo are off by almost $\frac{1}{32}$". This is an excellent reason to use a single rule throughout a project; if you use multiple rules and tape measures, you may introduce errors that can't be traced.

Measuring Tips

Pros know the importance of precision measurements and layout. Here are a couple of tricks that can help.

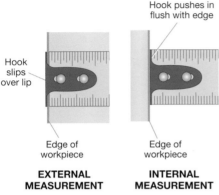

Hook pushes in flush with edge

Hook slips over lip

Edge of workpiece

Edge of workpiece

EXTERNAL MEASUREMENT

INTERNAL MEASUREMENT

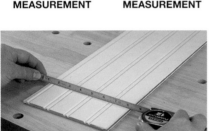

USING THE HOOK ON A TAPE MEASURE. The hook on a tape measure is designed to move freely back and forth for inside and outside measurements. On inside measurements it will slide in the exact thickness of the hook to provide an accurate measurement.

READ TAPE MEASURE AT 1" MARK. When taking outside measurements with a tape measure, try extending the hook end of the tape 1" past the edge of the workpiece. Take the reading and subtract an inch. This method eliminates any possible error caused by a bent or malfunctioning hook.

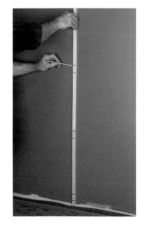

STORY STICK. Story sticks have been around since humans started building. A story stick is a scrap of lumber that's cut slightly longer than the work area. Then the location of horizontal trim pieces are accurately measured and drawn directly on the stick. To use a story stick, place it against a wall and transfer its marks onto the wall. Now you can move the stick along the wall, marking as you go without the need for measurements.

Using a Stud Finder

The fasteners that hold trim in place need a solid purchase—either a wall stud, ceiling joist, or top or bottom plate. So the first order of business for many trim jobs is to locate these. The problem is that they're all hidden under sheathing (typically drywall). The solution? An electronic stud finder.

1 SCAN FOR STUDS. To use an electronic stud finder, position the finder on the wall, depress the ON button, and gently slide it along, as shown in the top photo. The stud finder will have a visual display to indicate that you've found the edges of the stud; some finders also have an audible indicator.

2 MARK THE STUDS. When you find the edge of the stud, stop and mark its location (middle photo). Then lift up the finder, place it about 3" away from your mark, and slide back toward the mark to identify the other edge. Mark it, and then draw an X between the lines to indicate the center of the stud.

PRO TIP

Easy-to-Remove Marks

The problem with marking on walls is that you have to go back and remove the marks (sometimes with limited success). To eliminate this problem, pros run a strip (or strips) of masking tape along the wall (they use painter's tape, since it won't damage the surface). Then they mark the tape and when done, peel it off, leaving an unmarked wall.

Marking Reveals

The casing that runs around the perimeter of a window or door opening is typically offset from the jamb. This offset—often called a reveal—does a couple of things. It allows for easier installation and provides a shadow line for visual interest. Additionally, it helps conceal any variations in the trim and/or the jamb. There are two ways to mark a reveal: with a combination square and with a finger gauge (see below).

COMBINATION SQUARE. To mark a reveal with a combination square (page 40), set the blade the desired distance from the edge of the jamb (about ⅛" or so) and lock it in place. Then butt the head up against the jamb. Position a pencil in the small notch centered in the end of the metal rule. Then, while holding it firmly in place, slide the square along the edge of the jamb, keeping steady pressure on the head.

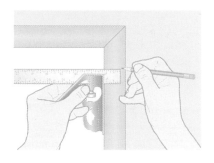

FINGER GAUGE. If you don't have a combination square, you can use your fingers as a simple gauge. Hold the pencil in your hand, and use your fingers as a stop to position the pencil so it extends out the desired amount. Then slide your hand down the jamb to mark the reveal.

Measuring Angles

We've already established that most adjoining surfaces in a home are not level, plumb, or square. This is what makes cutting miter joints (page 20) such a challenge. To miter parts accurately, you need to know the true angle of the intersecting surfaces. There are a few ways to do this: with an angle gauge, with a sliding bevel, with a compass, or with a digital protractor.

ANGLE GAUGE. One of the quickest ways to identify an angle is to use an angle gauge. To use an angle gauge, press the blades of the gauge firmly against the surface to be measured and then tighten the tension nut. The angle can then be read directly off the gauge.

SLIDING BEVEL. If you don't have an angle gauge, you can determine the angle with a sliding bevel.

1 DRAW PARALLEL LINES. Start by drawing a line parallel to each surface, as shown.

2 IDENTIFY THE ANGLE. Then press the stock of the sliding bevel against one surface and pivot the blade until it bisects the parallel lines, as shown. Tighten the blade lock and either use this to set your miter saw to the exact angle or use a protractor to measure the angle and use this to set your saw.

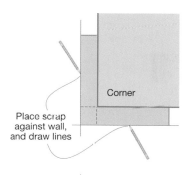

Place scrap against wall, and draw lines

Corner

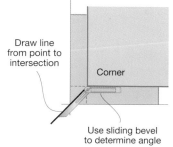

Draw line from point to intersection

Corner

Use sliding bevel to determine angle

WITH A COMPASS. Another way to bisect an angle is to use a compass as illustrated in the drawing below.

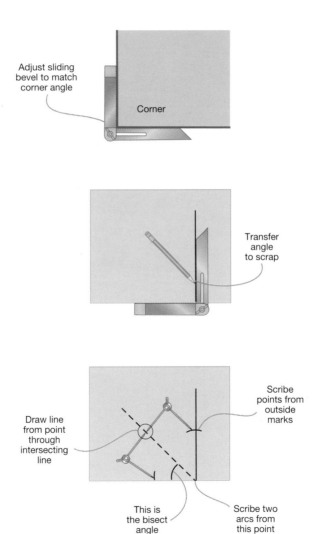

Adjust sliding bevel to match corner angle

Corner

Transfer angle to scrap

Draw line from point through intersecting line

Scribe points from outside marks

This is the bisect angle

Scribe two arcs from this point

Using a Digital Protractor

If you're planning to install a lot of trim, consider buying a digital protractor, like the Bosch DWM40L. A digital protractor takes all the guesswork and computations out of installing trim. Simply measure the wall angle with the protractor, and it will compute the miter and bevel settings for your saw. Then just set up your saw with the angles provided by the protractor and make your cut to get a perfect fit.

1 MEASURE THE ANGLE. To use a digital protractor, open up the legs of the protractor and butt each leg flat against the wall surfaces and read the digital display. It can also be used to determine the spring angle of trim.

2 STORE THE ANGLE. If the protractor has a store or hold function, press the hold button to store the wall angle so that you can compute the saw settings.

3 CALCULATE THE CUT. For the protractor shown here, press the BV/MT button and the display will show the value of the miter setting. Make a note of this or adjust your saw miter to this angle. Then press the BV/MT button

to show the bevel angle. Make a note of this or adjust your blade tilt to this angle. Your saw is now set to cut the trim.

Removing Trim

Trim jobs often begin with removing old trim. Here's how to do it without damaging your walls.

1 CUT THE CAULK. If caulk was used to conceal gaps between the trim and the wall, start by cutting through the caulk with a utility knife. When dry, caulk can form an incredibly strong bond. If you don't cut through the caulk before pulling off the trim, you'll probably pull off part of the wall covering along with the trim.

2 PRY OFF THE TRIM. With the caulk cut, you can pry off the trim. To prevent a prybar or putty knife from damaging the wall, slip another putty knife between the trim and the wall. Then insert either a stiff-blade putty knife or a prybar between the first putty knife and the trim. Now gently pry the trim carefully away from the wall. If you're planning on reusing the trim, see page 69.

3 REMOVE EXCESS CAULK. Now go back and scrape off any caulk residue on the wall with a stiff-blade putty knife.

Removing Nails

Virtually every trim job you tackle involves removing nails—either from walls and ceilings or from the old trim. Which tools you'll use for this will depend on what you've got on hand, the location of the nail, and its size.

PRYBAR. For extra-stubborn nails, nothing beats a prybar. Slip the claws of the hooked end of a bar around the nail head and pull back on the bar to pull the nail free. Since a prybar exerts a lot of pressure at the fulcrum point, it's best to slip a putty knife under its head (as shown) to protect the wall covering.

CAT'S PAW. A cat's paw can remove stubborn or hard-to-reach nails, or those that have been driven flush with the surface of the trim. You can use it like a prybar (as shown) for stubborn or hard-to-reach nails. For flush nails, position the claws directly behind the nail. Hold it at an angle and strike the head with a hammer to drive the claws under the head of the nail. Then you can pry the nail up and out.

PRO TIP

Recycling Trim

If you're planning on reusing any of the trim you've removed, don't pound the nails out through the face of the molding. This will split the wood, creating a larger hole to fill. Instead, pull the nails out from behind with a pair of locking pliers or end nippers (as shown). This creates a much smaller hole to fill.

TRIM KNOW-HOW

Cutting Trim by Hand

Traditionally, trim was cut to length with a hand saw and a miter box. Miter saws are basically long backsaws with crosscut teeth designed to fit in a miter box. The box holds the trim piece and positions the saw at the desired cut angle. Manufactured miter boxes have fixed slots at 22½, 45, and 90 degrees. But these have been replaced in many home workshops with a power miter saw (page 71). Not only do power miter saws make effortless cuts, but they also adjust to any angle and can make compound cuts with ease. Still, for small trim jobs, a miter saw and box can work fine.

1 MEASURE AND MARK. To use a miter saw and box, start by measuring and marking your trim piece to the desired length.

2 CLAMP AND CUT. Next, position the trim in the box so your mark aligns with the desired angle slot, and clamp it in place. On the miter box shown here, a series of cam-like clamps (black vertical rods) hold the workpiece firmly in place.

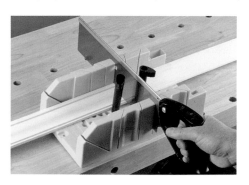

To cut the trim, take a couple of light strokes backwards to create a starting kerf, then stroke forward until you've cut through the trim.

Cutting Trim with a Power Miter Saw

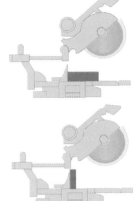

Modern compound miter saws can tilt and pivot their blades. And new technology (like laser guides) permits more accurate cuts. Sliding compound miter saws (page 72) increase cutting capacities. There are three ways to position trim for cutting on a miter saw: flat or on edge (see below) or angled (see pages 78–79).

FLAT VERSUS ON EDGE. You'll get more accurate cuts when you place a workpiece flat on the saw table, as this provides the most stable platform for the workpiece. For shorter workpieces, it's often more convenient to miter-cut molding (like baseboard) on edge than it is to lay it flat and angle the saw for a bevel cut.

MITER SAW SAFETY

1. Secure the miter saw to a stable supporting surface.
2. Use correct size and type blades for your saw.
3. Always use the blade guard.
4. Never start the saw with the blade against the workpiece.
5. Keep arms, hands, and fingers away from the blade.
6. Allow the motor to come to full speed before making the cut.
7. Do not cut small pieces.
8. Never perform freehand operations.
9. Properly support long or wide workpieces.
10. Don't let anyone stand behind the saw.
11. Lock the bevel, miter, and fence before using the saw.
12. Read the instruction manual before operating your saw.
13. Wear eye and hearing protection.
14. Keep your work area clean.
15. Always disconnect the power cord before making any adjustments or attaching any accessories.

TRIM KNOW-HOW

1 ALIGN WITH MARK.

With the miter saw adjusted to the desired angle, slide the trim over until your mark or line is aligned with the blade. If your saw is equipped with a laser guide, this is a snap.

2 MAKE THE CUT.

With the mark aligned, turn on the miter saw and pivot the blade down to make the cut. If you're using a sliding compound miter saw, use the method described below to cut the trim.

SLIDING COMPOUND SAW. Here's how to make a cut with a sliding compound miter saw. Start by pulling the carriage toward you until the blade is completely past the edge of the workpiece. Then lower the blade as shown. Next, push the saw back to its starting position while keeping the blade down, as shown. The teeth will push the workpiece firmly into the fence as the blade cuts, helping you cut accurately.

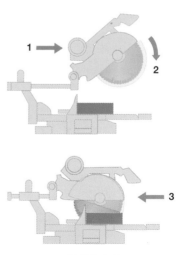

SLIDE CUT

Cutting Miters

As with straight crosscuts, you have two positioning options for trim when cutting miters: flat and on edge. Similarly, which position you choose will depend on personal preference and the size of the trim. Narrow trim can be cut on edge, while wider pieces will need to be mitered lying flat.

1 POSITION AND CLAMP. One of the most reliable ways to ensure an accurate miter is to clamp your trim in place before making a cut. Even the best crosscut blades have a tendency to pull the trim slightly as they cut into the wood, producing an inaccurate cut. Use either the built-in hold-down clamps (as shown here) or a shop clamp to securely lock the trim against the table or fence.

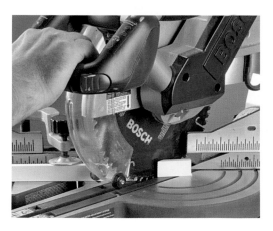

2 CUT THE MITER. To cut the miter, just pivot the blade down into the workpiece. For several tips on making accurate cuts, see pages 74–75.

Mitering Tips

Try these suggestions for more ways on improving trim cut accuracy.

PREVENT CREEP. Creep is caused by a blade's tendency to pull or push a workpiece when it makes a cut. One way to reduce creep is to temporarily attach sandpaper to the saw table and/or fence. The grit

of the sandpaper "grabs" the trim and keeps it from shifting. Self-adhesive sandpaper works great for this, or you can attach standard sandpaper with rubber cement.

USE BUILT-IN CLAMPS. Miter cuts really tend to pull or push trim during a cut; a compound cut does this even more, since you're really cutting through thicker stock. That's why it's so important to use a hold-down to keep the trim locked in place.

SUPPORT THE WORKPIECE. Before you make a cut, make sure the trim is fully supported, especially on long pieces: These can tilt and produce a mis-cut, or personal injury. Depending on your trim's length, this may be as simple as extending the built-in sliding table extensions of your saw, as shown. For longer trim, consider using a roller stand.

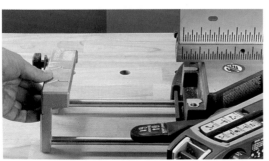

USE SHIMS TO TWEAK ANGLES.

You'll often find that even with careful layout and positioning, you still need to tweak the angle of a cut. One way to do this is to shim the trim instead of trying to adjust the saw. A standard playing card works

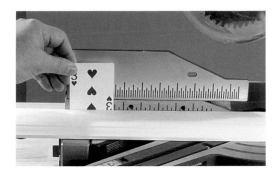

great for this. Just slip it between the trim and the fence to push the end of the trim out slightly and vary the cutting angle by just a hair. If that isn't enough, just add another card.

USE A STOP.

The best way to make accurate repeat cuts is to use a stop. Stops can be built-in, after-market, or shop-made. For short pieces, clamp a scrap to the fence, as shown in the middle photo. Many miter saws have built-in stops like the one shown in the bottom photo. These too have limited use, since they can usually extend out only as far as the table extensions. For longer

work you'll need a fence extension like the one built into the Bosch miter saw stand, shown on page 50.

Cutting Compound Miters

Acompound cut is any crosscut where the blade is set at an angle to both the saw table and the saw fence. Because you're angling both the table and the blade, accurate compound cuts are a challenge to set up. That's why it's important to make a set of test cuts on scrap before committing to cut your trim—a certain amount of trial and error is required with most compound cuts.

1 PIVOT THE FENCE. Instead of setting both angles at once, it's best to set one angle first—the miter angle. Once it's set to the desired angle, make a test cut on scrap and measure the cut to verify that the angle is correct.

2 TILT THE BLADE. Now tilt the blade to the desired angle and make a test cut on scrap. Measure the bevel angle before cutting the first scrap that you laid out the compound angle on—it's a lot easier tackling the angles one at a time.

3 MAKE THE CUT. Cutting trim at an angle means you're effectively cutting a wider board; cutting trim at a bevel means you're cutting a thicker board. So when you make a compound cut, you're effectively cutting into wider, thicker trim. That's why it's best to slow down the feed rate of the blade as it passes into and through the workpiece. If you cut too fast, odds are that the blade will deflect slightly, resulting in an inaccurate cut.

Cutting Crown Molding

There are two methods for cutting crown: flat on the table or angled against the fence. The crown we're cutting here has a 52/38-degree spring (see page 12). To cut crown flat, see the chart on page 78 for miter and bevel settings.

Cutting crown with a jig

If you want to cut crown angled against the fence and your saw doesn't have built-in stops (see page 79), consider using a crown molding jig. Crown molding jigs hold trim at the correct angle for the cut. The crown jig we're using here is manufactured by Bench Dog Tools (www.benchdog.com).

1 POSITION THE CROWN. Begin by placing your crown in the jig. On most jigs, there's an adjustable bar that you can slide back and forth to position the molding. You want the top flat flush with the jig's fence and the bottom flat flush with the jig's base.

2 MAKE THE CUT. Now adjust the saw to the desired angle (45 degrees, in our case), slide the jig over so your mark aligns with the blade, and make the cut.

Common Crown Angles and Saw Settings

ANGLE BETWEEN WALLS	52/38-DEGREE MOLDING		45/45-DEGREE MOLDING	
	MITER SETTING	BEVEL SETTING	MITER SETTING	BEVEL SETTING
70 degrees	41.3 degrees	40.2 degrees	45.2 degrees	35.6 degrees
71 degrees	40.8 degrees	39.9 degrees	44.7 degrees	35.2 degrees
72 degrees	40.3 degrees	39.6 degrees	44.2 degrees	34.9 degrees
73 degrees	39.7 degrees	39.3 degrees	43.7 degrees	34.6 degrees
74 degrees	39.2 degrees	39.0 degrees	43.1 degrees	34.3 degrees
75 degrees	38.7 degrees	38.6 degrees	42.6 degrees	34.1 degrees
76 degrees	38.2 degrees	38.3 degrees	42.1 degrees	33.8 degrees
77 degrees	37.7 degrees	38.0 degrees	41.6 degrees	33.6 degrees
78 degrees	37.2 degrees	37.7 degrees	41.1 degrees	33.3 degrees
79 degrees	36.7 degrees	37.4 degrees	40.6 degrees	33.0 degrees
80 degrees	36.2 degrees	37.1 degrees	40.1 degrees	32.8 degrees
81 degrees	35.8 degrees	36.8 degrees	39.6 degrees	32.5 degrees
82 degrees	35.3 degrees	36.5 degrees	39.1 degrees	32.2 degrees
83 degrees	34.8 degrees	36.1 degrees	38.6 degrees	32.0 degrees
84 degrees	34.3 degrees	35.8 degrees	38.1 degrees	31.7 degrees
85 degrees	33.9 degrees	35.5 degrees	37.6 degrees	31.4 degrees
86 degrees	33.4 degrees	35.2 degrees	37.1 degrees	31.1 degrees
87 degrees	32.9 degrees	34.8 degrees	36.6 degrees	30.8 degrees
88 degrees	32.5 degrees	34.5 degrees	36.2 degrees	30.5 degrees
89 degrees	32.0 degrees	34.2 degrees	35.7 degrees	30.2 degrees
90 degrees	31.6 degrees	33.8 degrees	35.2 degrees	30.0 degrees
91 degrees	31.1 degrees	33.5 degrees	34.8 degrees	29.7 degrees
92 degrees	30.7 degrees	33.1 degrees	34.4 degrees	29.4 degrees
93 degrees	30.3 degrees	32.8 degrees	33.8 degrees	29.1 degrees
94 degrees	29.8 degrees	32.5 degrees	33.4 degrees	28.8 degrees
95 degrees	29.4 degrees	32.1 degrees	32.9 degrees	28.5 degrees
96 degrees	29.0 degrees	31.8 degrees	32.4 degrees	28.2 degrees
97 degrees	28.5 degrees	31.4 degrees	32.0 degrees	27.9 degrees
98 degrees	28.1 degrees	31.1 degrees	31.5 degrees	27.6 degrees
99 degrees	27.7 degrees	30.7 degrees	31.3 degrees	27.3 degrees
100 degrees	27.3 degrees	30.4 degrees	30.6 degrees	27.0 degrees
101 degrees	26.9 degrees	30.0 degrees	30.2 degrees	26.7 degrees
102 degrees	26.5 degrees	29.7 degrees	29.8 degrees	26.4 degrees
103 degrees	26.0 degrees	29.3 degrees	29.3 degrees	26.1 degrees
104 degrees	25.6 degrees	29.0 degrees	28.9 degrees	25.8 degrees
105 degrees	25.2 degrees	28.6 degrees	28.4 degrees	25.5 degrees
106 degrees	24.8 degrees	28.3 degrees	28.0 degrees	25.1 degrees
107 degrees	24.5 degrees	27.9 degrees	27.6 degrees	24.8 degrees
108 degrees	24.1 degrees	27.6 degrees	27.1 degrees	24.5 degrees
109 degrees	23.7 degrees	27.2 degrees	26.7 degrees	24.2 degrees
110 degrees	23.3 degrees	26.8 degrees	26.3 degrees	23.9 degrees

Using a built-in stop

The standard procedure for cutting crown molding is "upside down and backwards." That is, when you position the workpiece on the miter saw, you want to turn the molding upside down and flip it end-for-end. If your saw has a built-in stop (like the Bosch saw shown here), cutting crown is greatly simplified.

1 ADJUST THE STOP. To adjust a built-in stop, slide the stop in or out until the top flat rests flush with the fence and the bottom flat is flush with the saw table.

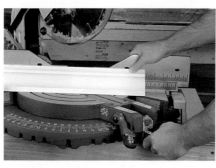

2 POSITION THE WORKPIECE. Once the crown molding is angled correctly, all you have to do with 45/45-spring molding is adjust the miter table to 45 degrees and lock it in place.

3 MAKE THE CUT. Make sure to hold the crown in place with one hand or a hold-down to keep it from shifting as you make your cut.

TRIM KNOW-HOW

Coping Trim

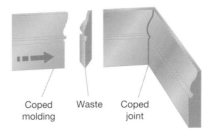

Coped molding   Waste   Coped joint

When trim carpenters want trim to intersect at inside corners without the gaps normally associated with miter joints, they cope the joint. There are two halves to a coped joint. On one half, the molding profile is left intact and butted into a corner. The second half is coped to fit the profile of the first molding butted into the corner.

1 EXPOSE THE COPE. To make a coped joint, start by butting one trim piece into a corner. Next you'll expose the cope on the second piece. This is done on the miter saw. The idea here is to cut the end as if you were cutting an inside miter. The miter cut will expose the wood that needs to be removed to fit against the matching profile of the first trim piece.

2 **REMOVE THE WASTE.** Before you cut into the waste portion of the trim, it's a good idea to make a series of relief cuts first where the profile changes direction. Then go back and cut out the waste. What you're after here is a 45-degree cut in the opposite direction of the exposed miter.

3 **FINE-TUNE AS NEEDED.** After you've removed the exposed waste, test the fit. Note any areas where there are gaps, and mark these. For large gaps, go back to the coping saw and remove most of the waste. For small gaps, use a file to fine-tune the profile. Test the fit frequently, and continue tuning until the coped part fits perfectly against the profile.

QUICK FIX

Using the EasyCoper

Want a quick way to cope molding? Try the EasyCoper as described on page 49. Start by exposing the cope as if you were using a coping saw. Then insert the trim in the EasyCoper and shift it to the opposite side from where you're cutting. Make a series of relief cuts, then cut into the waste area, following the pro-file until you're about halfway. Shift the trim to the opposite side and finish the cut.

TRIM KNOW-HOW

Making a Mitered Return

Occasionally, trim needs to stop before reaching a corner. Instead of cutting a butt or angled miter and leaving the end grain exposed, pros make a mitered return. This way no end grain is left exposed and the trim is neatly terminated.

1 MITER THE TRIM. To make a mitered return, begin by back-mitering the trim piece to leave a 45-degree end that faces the wall.

2 MAKE THE RETURN. Now miter-cut a scrap of trim at 45 degrees opposite that of the back-cut miter you just made. Hold this end up against the molding and mark it to length. Then use a hand saw to cut the return to length so the 90-degree end butts up against the wall.

3 ATTACH THE RETURN. Apply glue to the mitered ends of the trim and return and press the pieces together. Apply masking tape across the joint to hold the pieces in place until the glue dries.

Fine-Tuning Trim

Depending on how out-of-square a room is, you may need to fine-tune a piece of trim to get it to fit properly. A simple way to mark a piece of trim for this is to "scribe" it; see below.

SCRIBING TO FIT. To scribe a piece of trim, place the trim where it will be installed. Then set a compass to slightly wider than the largest gap between the trim and the surface you're scribing. Next, place the compass point on the surface with the pencil tip resting on the trim. Now slide the compass along the surface. The pencil will scribe the unevenness of the surface directly onto the trim so you can cut it to match.

USING A BLOCK PLANE. A block plane is the perfect tool to fine-tune trim. You can either slide the plane over the trim to make a cut, or pass the trim over the block plane. Take light cuts and check the fit often.

P R O   T I P

Shooting Board

A shooting board is a shop-made jig that holds trim so you can accurately trim it with a plane. A shooting board has a two-step base; the top step sports one or more straight or angled cleats. The trim is held against one of the cleats, and the plane is laid on its side on the bottom step and passed over the end of the trim to cut it to size.

Attaching Trim by Hand

Before the advent of air nailers (pages 86–87), trim was attached with a hammer and a nail set.

HAMMER CLOSE TO THE SURFACE. The trick to attaching trim by hand is knowing when to stop driving the nail with a hammer. The closer you get to the surface, the greater the chance of dinging it with the hammer. To eliminate dings, stop when the head of the brad or nail nears the surface.

THEN SWITCH TO A NAIL SET. Reach for a nail set to finish the job and drive the nail below the surface. Nail sets are available is various tip diameters; choose one that is closest to the diameter of the nail head. Place the set on the nail and tap it to set the nail below the surface.

FILL THE HOLE. When all the trim is installed and the nails have been set, go back and fill the holes with putty. See page 36 for more on putty types and page 96 for more on applying putty.

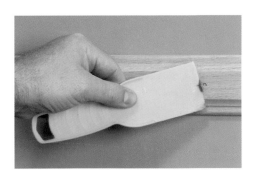

Nailing Secrets

Many trim carpenters still use a hammer to install trim. Here a few tricks of the trade to achieve professional results.

LOCK-NAILING.
Miter joints have a tendency to open and close over time. One way to lock them in place is to "lock-nail" the joints together. Use a finish nail for this, and make sure to stay away from the front edge to keep from splitting the trim.

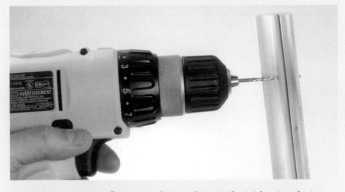

PRE-DRILLING. Seasoned pros know that trim tends to split when you nail near an edge—particularly a thin edge—and near the end of a piece. To prevent this, trim carpenters drill a pilot hole in the trim piece before nailing it in place. Another time-honored trick is to blunt the tip of the nail before using it—this creates a dull point which lessens the chance of a split.

Attaching Trim with Air Tools

I f you've never used an air nailer, you're in for a treat: no more dings, no more nail sets. An air nailer drives and sets a nail with the pull of a trigger.

POSITION AND SHOOT. To use an air nailer, load it with the appropriate fastener and put on eye protection (see page 87 for more on air nailer safety). Hold the trim in place and press the tip of the nailer into the trim until the tip engages the safety. Then squeeze the trigger to fire a fastener. What could be easier?

PIN ANY MITERS. When the trim you're attaching is mitered, it's best to "pin" the miter joint with brads, as shown in the bottom photo. This is easily accomplished with an air nailer, but can be a real challenge with a hammer and nail.

ALWAYS DISCONNECT HOSE WHEN RELOADING. The number one safety rule for using an air nailer is to always wear eye protection. The number two rule is to always disconnect the air hose before loading or removing fasteners. This takes only a second and can help prevent a nasty accident.

AIR TOOL SAFETY

1. Wear eye and hearing protection.

2. Always disconnect the air hose when loading or removing fasteners or servicing the tool.

3. Make sure the compressor is supplying the recommended pressure to the tool; do not exceed the maximum pressure of the tool.

4. Never carry an air nailer with your finger on the trigger.

5. Never point an air nailer at yourself or another person.

6. Make sure you have plenty of hose clearance to prevent tripping.

7. Use only the fasteners recommended by the tool maker; it's best to use their brand.

8. Inspect and lubricate the tool periodically according the manufacturer's instructions.

9. Keep your fingers away from the tip of the gun; wood can make fasteners deflect, causing an injury.

10. Make sure there's no one behind your work area. Air nailers are powerful enough to shoot fasteners completely through some materials.

Using a Biscuit Jointer

A biscuit jointer (also called a plate jointer) cuts half-moon-shaped slots. Pairs of opposing slots are cut in pieces to be joined, and a football-shaped compressed wood biscuit is inserted in the slots. Not only does this align the surfaces, but also, when glue is applied, the biscuit swells to lock the pieces together. Biscuit joints are great for strengthening butt joints (as shown here), as well as miter and scarf joints.

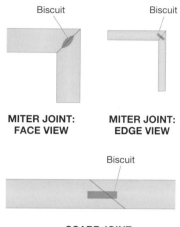

MITER JOINT: FACE VIEW

MITER JOINT: EDGE VIEW

SCARF JOINT: EDGE VIEW

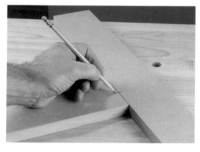

1 MAKE REGISTRATION MARKS. To use a biscuit jointer, set the parts to be joined in their final position. Then make a registration mark across the joint line, centered where the biscuit will be inserted.

2 ALIGN THE TOOL WITH THE MARKS. Now adjust the biscuit jointer as needed for the correct size biscuit (usually either a #10 or #20) and set it to cut a slot centered on the thickness of the workpiece. Butt the end of the biscuit jointer against the edge or end of a workpiece and align its center-mark (red vertical line on base of face) with your registration mark.

3 **PUSH IN TO CUT JOINT.** Now turn on the biscuit jointer and push it into the workpiece to cut the slot. Springs in the body of the tool will automatically back it out of the cut once the slot has been cut. Repeat for the other workpiece.

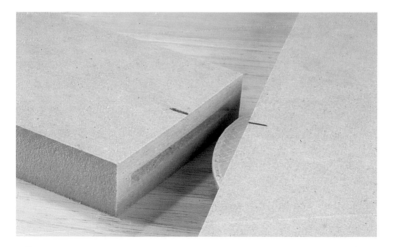

4 **ADD THE BISCUIT.** Before applying glue, check the fit of the biscuit and make sure the registration marks align. If all looks good, separate the parts, remove the biscuit, apply glue, and reassemble. If possible, add clamps to keep the parts from shifting while the glue dries.

Alternative Techniques for Attaching Trim

For most trim jobs, the trim will be attached to wall studs or ceiling joists. There are exceptions, though, including: using foundation trim, backer panels, or blocks and strips, or when framing members aren't present.

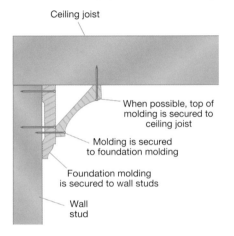

Ceiling joist

When possible, top of molding is secured to ceiling joist

Molding is secured to foundation molding

Foundation molding is secured to wall studs

Wall stud

Foundation trim

Foundation trim is commonly used for ceiling trim—particularly if the trim is large or heavy, as is often the case with crown molding. The idea is to attach a piece of flat trim to the wall studs and then attach the ceiling trim to the foundation trim. This lets you fasten the ceiling trim to the foundation anywhere along its length instead of just at the wall studs.

Backer panels

A version of foundation trim is backer panels. They serve the same purpose, just on a larger scale. Backer panels are often used when installing flat or panel wainscoting (as described on pages 136–141).

1 ATTACH THE PANEL. Like foundation trim, backer panels are secured to the wall studs. The drywall may or may not be removed first.

2 INSTALL THE TRIM.
Once the panel is secure, the trim can be attached anywhere along the backer panel.

Backer blocks

Backer blocks are individual versions of foundation trim. Instead of long strips, smaller angled blocks are used. This is a popular method for attaching crown, since an angled block provides a larger nailing surface and also supports the trim at the correct angle (see page 155).

Backer strips

Backer strips are a hybrid of backer blocks and foundation trim. Like foundation trim, they are strips. And like backer blocks, they are not left exposed—the trim covers them completely. Backer strips are often used when installing wainscoting (pages 126–132).

1 ATTACH THE STRIPS. Like panels,
backer strips attach to wall studs. Here again, the wall covering may or may not be removed first (as shown here).

2 ATTACH THE TRIM.
With the strips in place, the trim can now be fastened to them.

No framing members

Occasionally you'll find areas where there are no framing members for attaching trim. A common example of this is the walls that run parallel to ceiling joists, as illustrated in the bottom drawing. There are two things you can do to help hold trim in place in these situations: Use construction adhesive, and cross-nail.

USE CONSTRUCTION ADHESIVE. Apply a liberal amount of quality construction adhesive to the trim (shown here are wall frames) before pressing the trim in place. Then either cross-nail (see below) or apply strips of masking tape to hold the trim in place until the adhesive sets up.

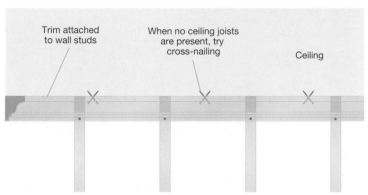

Trim attached to wall studs

When no ceiling joists are present, try cross-nailing

Ceiling

CROSS-NAILING. Cross-nailing is simply driving in fasteners at opposing angles. This effectively pinches the trim against the wall at the cross-nail locations to help hold it in place.

Sealing Exterior Trim

Trim installed around windows and doors conceals the gaps between the rough framing and the window or door jamb. On exterior windows and doors, these gaps should always be filled to prevent drafts, usually with fiberglass or foam insulation. And the perimeter edges of the trim should also be sealed with caulk; see below.

FIBERGLASS INSULATION. Fiberglass insulation does an admirable job of preventing drafts. The big thing to remember is that insulation works by trapping air. If you pack it in too tight, the insulation can't do its job.

INSULATING FOAM. Insulating foam also works well. Just remember to use the minimal-expanding type designed for windows and doors. If you use too much, the expansion can bow the jamb, resulting in a sticky window or door.

CAULK AROUND TRIM. Once your exterior trim is in place, take the time to apply 100% silicone caulk around the perimeter of the trim to further seal against drafts.

Making Your Own Molding

There are a couple of reasons why you may want to make your own molding: You're trying to match existing trim and can't find a match, or you just want to be creative. If you've got a couple of wood-working tools—a router and a power

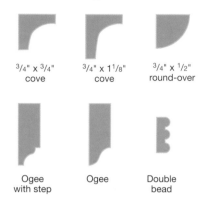

³/₄" x ³/₄"
cove

³/₄" x 1¹/₈"
cove

³/₄" x ¹/₂"
round-over

Ogee
with step

Ogee

Double
bead

saw—it's easy to make your own. In addition to simple routed edge profiles (top drawing), you can also create complex profiles by gluing up profiled strips, as described on page 95. **Safety note:** It's unsafe to rout narrow strips because they can flex and snap while routing. A safer way to do this is described below.

1 ROUT PROFILES ON WIDE STOCK.
To make your own molding, begin with a wide piece of wood that's cut a bit longer than your finished trim length. Then use a router fitted with the desired profile to rout a profile along its full length. Although the full profile is shown here, it's best to take multiple passes—usually three light cuts to sneak up on the final profile. This is especially important when working with thin stock.

2 **RIP MOLDING TO WIDTH.** Once you've profiled the edge, use a power saw to rip the trim to the desired width. A table saw is the best tool for this job, but you can also cut the trim with a circular saw fitted with a rip fence.

TRIM KNOW-HOW

Finishing Trim

How you finish trim will depend on whether it's being painted or finished with a clear topcoat. Nail holes will also have to be dealt with, and *when* you fill them will depend on the type of finish.

Filling nail holes with putty

There are two types of putty for filling nail holes: hardening and nonhardening. Nonhardening putty is applied in a similar manner as wax crayons are, as described on page 97.

1 APPLY THE PUTTY. To apply hardening putty, press it into the nail hole with a putty knife. Leave the putty a bit proud of the surface: You'll sand it flush in the next step.

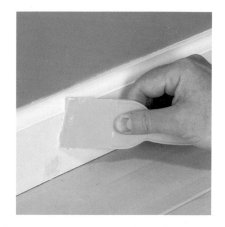

2 SAND IT FLUSH. Let the putty dry the recommended time and then sand it flush with the surface. Wrap a piece of open-coat sandpaper around a scrap block, as shown. The block prevents you from dishing into the putty as you sand.

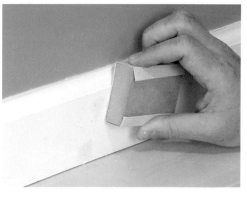

Filling nail holes with wax

Trim carpenters often pre-finish solid wood trim with a clear topcoat before installing it (see page 99). Then they use wax crayons to fill the nail holes once the trim is in place, as described below.

1 MATCH THE COLOR. To use a wax crayon, start by matching a crayon to the trim. Don't go by the name on the crayon (such as natural oak). Instead, buy a range of colors and return the ones you don't use.

2 APPLY THE WAX. Fill a nail hole by scrubbing the tip of the crayon over the hole. Overfill it slightly.

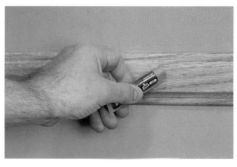

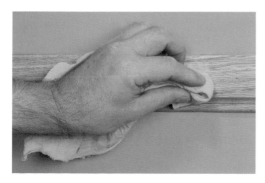

3 BUFF OFF EXCESS. Now use a clean, dry cloth to buff away any excess wax until it's flush with the surface of the trim.

Applying paint

Paint is one of the hardiest finishes you can apply to trim. It's always used on paint-grade or pre-primed trim to obscure any differences in the woods used to make the trim. For trim, choose satin, semi-gloss, or eggshell, as these finishes handle dirt and fingerprints better than flat paint.

1 MASK OFF SURFACES. To paint trim, start by protecting adjacent and surrounding surfaces by masking them off (top photo).

2 APPLY A PRIMER. Priming is an important step that ensures a good bond between the trim and the new paint (middle photo). It's formulated to make the old surface more "receptive" to the paint. Priming also seals damaged areas and hides stains.

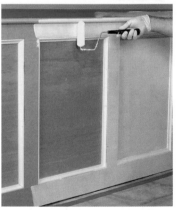

3 PAINT. Now you can paint the trim. A small roller like the one shown in the bottom photo will make quick work of large areas. Strips of trim are best painted with a foam brush.

Applying a clear finish

Wood (either stained or left natural) needs to be protected with a clear topcoat. No matter which topcoat you choose, it's best to apply multiple thin coats instead of a single heavy coat. By applying a thinner coat, there's less chance of finish building up in nooks and crannies, resulting in a sag or run. It's also a good idea to wait a couple of minutes after the finish has been brushed on and inspect it carefully with a light held at a low angle; you're looking for sags and runs. Remove any excess finish from your brush by wiping it on a clean cloth, and then go back and brush out any runs or sags that you've found.

P R O T I P

Pre-Finish Whenever Possible

Pros know that it's a real hassle to apply a finish to trim once it's installed. This is especially true for trim attached at or near the ceiling. Not only do you end up applying the finish on a ladder, it's too easy to get finish on adjacent walls and other surfaces. That's why whenever

possible, many pros pre-finish the trim before installation. Then all they have to do is go back and fill nail holes with wax or nonhardening putty and touch up the occasional spot.

5

Floor Trim

FLOOR TRIM DOESN'T GET the star billing of, say, crown molding or wainscoting, because its function is deceptively ordinary: offering a transition between floors and walls. But without floor trim, you'd see gaps between flooring and walls, and lose the visual interest it adds to a room. Wood and laminate flooring especially need trim, because they require a perimeter gap to allow for seasonal expansion and contraction. In this chapter, we'll show you how to install both one-piece and built-up baseboard and how to work around common obstacles.

One-Piece Baseboard

Although one-piece baseboard is the simplest baseboard to install, it can still add a lot to a room. Trim manufacturers offer a wide variety of profiles and sizes. Profiles range from a simple single roundover on top of the trim to complex, sculpted shapes that can run from the top to the bottom. Sizes vary from 2½" up to over 6" in width (height). Most baseboard comes in 8- or 16-foot lengths. As a general rule of thumb, use wider (taller) baseboard in rooms with higher ceilings. Also, if your floors are uneven and you can't level them, consider using foam trim—it's a lot more flexible than wood and can be gently shaped to fit your floor. Wood trim is machined with a recess on its back, as illustrated in the drawing above. This creates two nailing flats that allow the trim to lie flat on uneven walls.

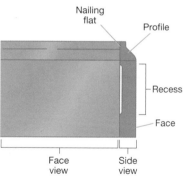

Nailing flat

Profile

Recess

Face

Face view

Side view

1 **CHECK FLOOR FOR LEVEL.**
Before you start cutting trim, take the time to check your floor for level. If you're installing new flooring, stop and level the floor. Self-leveling compounds are available at most home centers and make quick work of this job.

2 **LOCATE AND MARK STUDS.**
The next step is to locate the wall studs and mark their position. This is a simple task with an electronic stud finder, as described on page 63.

as described on page 63.

PRO TIP

Fine-Tuning Walls

Baseboard will lie flat on a wall only if the wall is flat and plumb. Trim carpenters know that drywall joint compound (often called mud) can build up at joints and especially at corners, and ruin the fit. That's why you'll often see them pull out a putty knife and scrape a section of a wall flat. This is much more efficient (and better looking) than installing trim on an uneven wall and filling in the resulting gap or gaps with caulk.

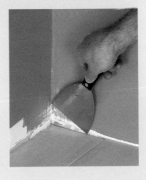

FLOOR TRIM

3 BUTT TRIM INTO CORNER.

With the walls prepped and the studs located, you can start installing trim. Measure the longest, most visible wall and cut a piece of baseboard to fit; butt it into the corners. (See page 11 for more on the typical sequence used to install perimeter trim.)

Corner Blocks

Don't want to mess around with coping or mitering your baseboard to fit the inside and outside corners of the room? Consider installing corner blocks. These pre-made profiled blocks are glued and nailed into the corners. The baseboard is cut straight on its ends and simply butts up against the corner blocks.

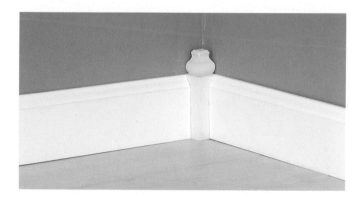

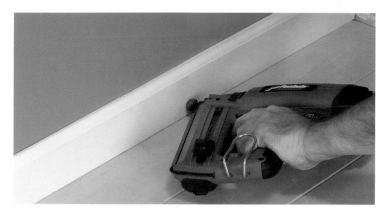

4 **SECURE THE BASEBOARD.** With the first piece of baseboard in place, check to see if there are any gaps between it and the flooring. If you find any, scribe and fine-tune the trim as described on page 83. When you're satisfied with the fit, secure the first piece, as described in the sidebar on page 108. Move on to the next piece, coping one end as described on pages 80–81, and work your way around the room until the perimeter is complete. If you encounter any obstructions (such as a heat register), see pages 110–113.

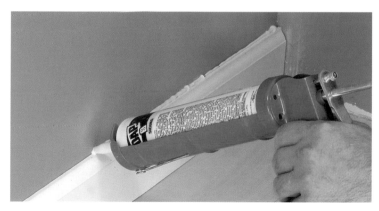

5 **FILL ANY GAPS.** Even if you fine-tune the fit of the baseboard, you may still end up with some gaps between the top of the baseboard and the wall covering. Fill these gaps with paintable latex caulk or caulk that's color-matched to the trim. Wipe off any excess immediately and when dry, paint if necessary. Finally, fill in all nail holes and apply a finish if desired (see pages 96–99.)

Built-Up Baseboard

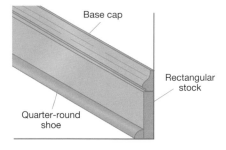

Base cap

Rectangular stock

Quarter-round shoe

Built-up baseboard is nothing more than multiple pieces of trim stacked on top of each other. It can be as simple as a base molding and a base cap, as illustrated in the drawing above, or as complex as a three- or four-layer buildup. Naturally, the more layers, the more exotic the profile.

Besides adding visual interest, there are two other reasons for using built-up baseboard. First, it can actually be less expensive to purchase simple trim and stack it up than to buy single heavily profiled trim. Second, if you're matching trim to an existing look, the trim may not be available. By mixing and matching stock trim, you can often create the look you're after. If you still can't match it this way, consider making your own trim (see pages 94–95.)

1 **LOCATE AND MARK THE STUDS.** To install built-up baseboard, start by checking your floor for level (page 102) and fine-tuning the walls if needed (page 103). Then locate the wall studs and mark their position. This is easy to do with an electronic stud finder, as described on page 63.

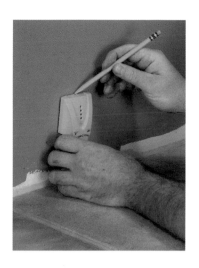

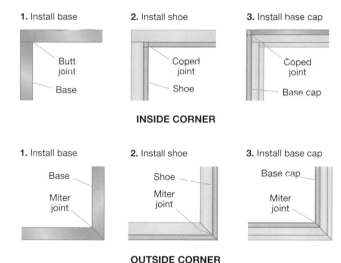

1. Install base

Butt joint

Base

2. Install shoe

Coped joint

Shoe

3. Install base cap

Coped joint

Base cap

INSIDE CORNER

1. Install base

Base

Miter joint

2. Install shoe

Shoe

Miter joint

3. Install base cap

Base cap

Miter joint

OUTSIDE CORNER

SEQUENCE OPTIONS. Installing built-up molding can be different than installing one-piece. For inside corners, when using a flat base molding that's not profiled, there's no need for coping—just butt it into the corners. Both the base cap and shoe should be coped as needed. Outside corners should be mitered. (See the sidebar on page 163 for a quick fix for open miters.)

2 INSTALL THE BASE PIECES.

Measure the longest, most visible wall, cut a piece of base to fit, and butt it into the corners. Attach the base as described in the sidebar below. Repeat for the remaining base pieces as you work your way around the room.

Securing Base Molding

Pros secure baseboard in two places: to the sill or sole plate and to the studs. Notice that this is where the nailing flats of the baseboard are located. If you drive a fastener into the middle of the trim—especially with a hammer followed by a nail set—you risk cracking the trim, since it's unsupported here. Pros will often use a heavier-gauge nail at the sill plate (typically 16-gauge), and a lighter-gauge nail at the more delicate top portion of the trim to prevent it from splitting.

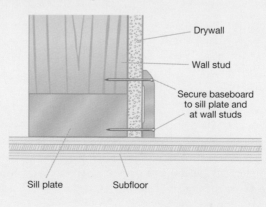

Drywall

Wall stud

Secure baseboard to sill plate and at wall studs

Sill plate   Subfloor

3 **INSTALL THE BASE CAP.** Now you can go back and install the base cap over the base trim. Some base cap is rabbeted on its bottom inside face to fit over the base trim. Others (like the base cap shown here) do not have a rabbet. Cut and fit the base cap as described on page 107, and secure it to the wall studs or base trim. You'll want to use a lighter-gauge fastener here to prevent splitting. Repeat for all the base cap.

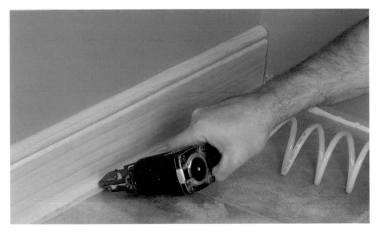

4 **INSTALL THE BASE SHOE.** All that's left is to cut and install the base shoe (if applicable). Just as with the base cap, you'll need to cope one end of each trim piece to fit over the profile of the previously installed piece. Fasten the shoe to the base piece or to the flooring (if possible).

Installing Baseboard around Obstacles

Installing baseboard—or any trim, for that matter—would be a snap if not for obstacles. The most common obstacles to work around are corners, both inside and outside. Other obstacles include built-in cabinets, electrical receptacles, heat registers, and surface-mounted electrical components.

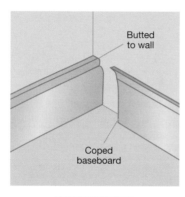

Butted to wall

Coped baseboard

INSIDE CORNER

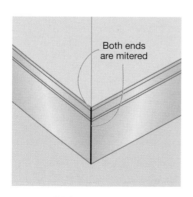

Both ends are mitered

OUSIDE CORNER

INSIDE CORNERS. An inside corner generally means the most amount of work and fuss when installing trim. That's because if you want a tight-fitting joint that doesn't open and close like a miter joint, you need to cope it. This entails cutting the end of one piece of baseboard to match the profile of the previously installed piece, as described on pages 80–81.

OUTSIDE CORNERS. Outside corners for baseboard are mitered. The only real challenge here is determining the angle of the corner, as it most likely isn't 90 degrees. See pages 65–66 for more on measuring angles and pages 73–75 for information on cutting miters.

WORKING AROUND CABINETS. Cabinets that are built into a wall and protrude out into a room can be handled as if they were a wall. Inside and outside corners are treated as described on page 110.

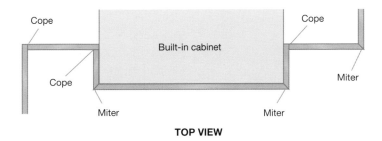

TOP VIEW

DEALING WITH RECEPTACLES. If your baseboard is wide (tall) and the receptacles in your home are low, you may need to make cutouts in the baseboard for receptacles. This same routine can be used to make cutouts in wainscoting (pages 126–141).

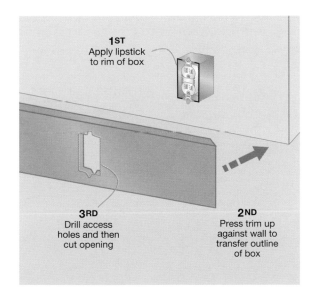

DEALING WITH REGISTERS. There are two methods for trimming around heat registers: wraparound and beveled. The wraparound method takes longer, but looks better. Beveling the ends of the trim is quick but not attractive.

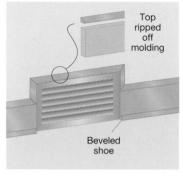

Top ripped off molding

Beveled shoe

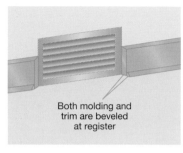

Both molding and trim are beveled at register

WRAPAROUND

BEVELED

SURFACE-MOUNTED ELECTRICAL. In homes where surface-mounted components are installed, you'll have to trim around these. Make sure to leave about ⅛" of clearance between the trim and the surface-mounted components to allow for seasonal wood movement.

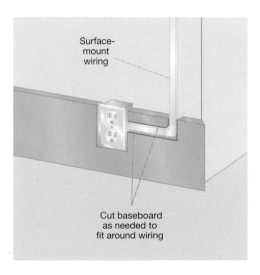

Surface-mount wiring

Cut baseboard as needed to fit around wiring

Working around Small Obstacles

★ Sometimes you'll encounter small obstacles that you need to work around when installing baseboard. The telephone jack shown here is one example. The only real challenge is accurately locating the cutout.

1 LOCATE OBSTACLE ON TRIM. Start by cutting your trim piece to fit the wall. Then position it against the obstacle as shown, and mark the position of the obstacle on the trim.

2 DEFINE ENDS OF CUT. Next, use a try square to extend the marks down the face of the trim, as shown.

3 LOCATE BOTTOM OF OBSTACLE. Measure up from the floor to locate the bottom of the obstacle (bottom left photo). Then transfer this to the trim piece and cut the opening with a hand saw or saber saw. Check the fit, and fine-tune as necessary.

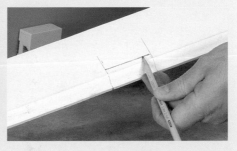

FLOOR TRIM

6

Wall Trim

WHILE FLOOR AND CEILING TRIM can add a lot to a room, it's the wall trim that really gets noticed. Accents like chair rail, picture/plate rail, wall frames, wainscoting, wall niches, and pilasters and columns catch the eye and add style to any room. In this chapter we'll show how to install each of these types of trim, and share tips for achieving professional results.

Installing Chair Rail

Chair rail not only protects your walls, but it can also be used to make a border between top and bottom wall expanses. Many homeowners like to treat the walls above and below the chair rail differently. So, you could paint the sections of the wall different colors, or wallpaper one section and paint the other. If you're planning to paint or wallpaper the room in question, do so now before installing the chair rail. You can buy pre-made chair rail in a variety of shapes, or you can make your own profile by building up the molding (see pages 94–95.)

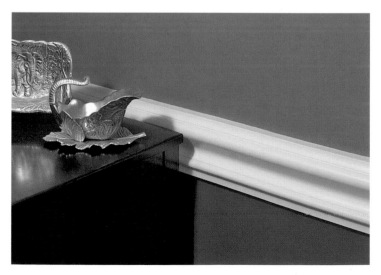

1 LOCATE THE RAIL. Chair rail is typically installed 32" to 36" above the floor. Measure this distance up from the floor and make a mark.

2 SNAP A CHALK LINE. Snap a level chalk line around the perimeter of the room as a reference line for installing the rail as shown in the bottom photo (or, use a laser level as described on page 118).

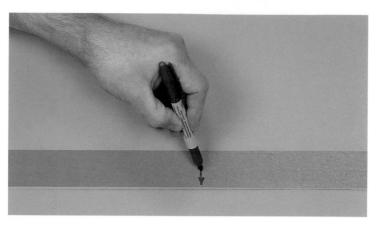

3 **LOCATE AND MARK THE STUDS.** After you've snapped your reference line, use an electronic stud finder to locate the wall studs as described on page 63. Use a pencil or marker to mark each stud (or see the sidebar on page 63 for a non-marring way to do this, as shown here).

Laser Level

You could snap a chalk line around the perimeter of your room to locate the chair rail, but then you'd have to remove the chalk from your walls. A better method is to use a laser level like the one shown here. The advantage of a laser level is that it shoots a perfectly level line along a wall—or around the perimeter of a room—without leaving any marks.

4 **BUTT TRIM INTO CORNER.** Cut your first piece of chair rail for the longest, most visible wall, and position it so its edge aligns with the chalk line you snapped earlier. Having a helper to hold the long strips of chair rail will make this job go much quicker.

5 **ATTACH THE TRIM TO STUDS.** When aligned, secure the chair rail to the wall at each stud location with finish nails. For fastening with air nailers, see pages 86–87; by hand, see pages 84–85.

6 SCARF AS NEEDED. For walls longer than your chair rail, you'll need to splice together pieces with a scarf joint, as described on page 21.

7 COPE AS NEEDED. If you're running the chair rail around the room, you'll need to cope the inside corners (pages 80–81) and miter the outside corners (see pages 73–75 for more on mitering). When all the chair rail is installed, go back and conceal any nail holes with putty, as described on pages 96–97.

Picture/Plate Rail

A plate rail or picture rail is a narrow shelf that's attached to the wall near the ceiling to display pictures or plates. It's usually a two-piece shelf made up of a top and a base. A groove is cut near the top front edge of the top to keep plates and pictures from accidentally sliding off the shelf; this also helps keep everything leaning at the same angle. The plate rail is mounted to the studs with screws since these shelves often support a lot of weight. An optional trim piece can be installed under the top to conceal the mounting screws.

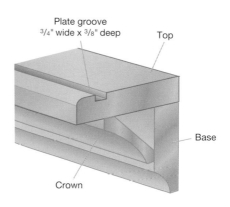

Plate groove
3/4" wide x 3/8" deep

Top

Base

Crown

3-PIECE PLATE RAIL

1 **LOCATE THE PLATE RAIL AND STUDS.** To install a plate/picture rail, start by measuring down the desired distance from the ceiling and make a mark. Then snap a chalk line around the perimeter of the room as a reference line for installing the rail (alternatively, use a laser level as described on page 118). Now use an electronic stud finder to locate the wall studs as described on page 63.

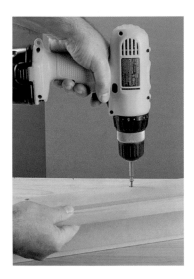

2 **BUILD THE RAIL.** Cut the base and top to size— the base shown here is $3\frac{1}{4}$" wide and the top is 4" wide. The groove is $\frac{3}{8}$" deep and is 1" in from the front edge. We also rounded over the top front edge of the top and cut a $\frac{1}{2}$" cove on the bottom face of the base. Glue and screw the top and base together, spacing the screws every 6" to 8". (It's also a good idea to apply a finish now before mounting the rail.)

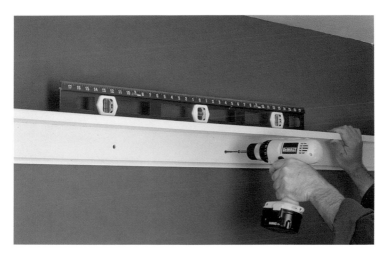

3 ATTACH THE RAIL. It's critical to mount the rail to the studs you've marked. So either measure these locations and transfer them to the plate rail, or hold the plate rail in place and mark their locations directly on the plate rail. Drill pilot holes and secure the rail to the studs with #10 × 3" woodscrews. Alternatively, you can use ¼" × 3" lag screws.

4 ADD THE MOLDING (OPTIONAL). If desired, cut pieces of crown or other angled trim and attach them under the top and to the base to conceal the mounting screws or bolts.

Installing Wall Frames

Wall frames are a great way to break up a large expanse of wall and to add visual interest to the wall. Wall frames can be large or small, and installed with or without a chair rail or other molding. The frames can be oriented vertically or horizontally and can be installed on the upper and/or lower portion of a wall. See the sidebar below on how to size your wall frames.

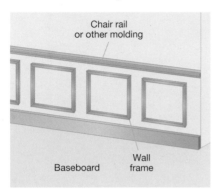

Chair rail or other molding

Baseboard

Wall frame

Golden Rectangle

In the most influential math book ever written, Euclid of Alexandria defined a proportion hailed as the most pleasing to the eye. (The complicated formula involved has to do with dividing a line into what Euclid called its "extreme and mean ratio." This ratio is equal to the ratio of 1 to 1.618.) The amazingly consistent occurrence of this ratio in nature has led scientists, mathematicians, and artists the world over to call it the "divine proportion" or "golden ratio." When used to construct a rectangle, it produces what's called a golden rectangle—the perfect solution for sizing your wall frames.

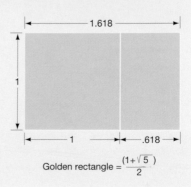

$$\text{Golden rectangle} = \frac{(1+\sqrt{5}\,)}{2}$$

1 ASSEMBLE THE FRAMES.

Once you've sized your frames, miter the ends as described on pages 73–75. Then make a simple jig to assemble them. The jig is just a pair of rectangles: a form and a base. The form

is cut to the exact inner dimension of the frame and attaches to the base. This automatically squares up the frame as it's assembled. Apply adhesive to the mitered ends, place the frame parts on the jig, and pin the ends together with brads.

2 APPLY AN ADHESIVE.

To attach a frame to a wall, begin by marking its height above the floor. Then snap a chalk line or use a laser level (page 42) to locate the top of the frames. Now mark the studs. Apply construction adhesive to the back of the frame and align its top edge with the chalk or laser line.

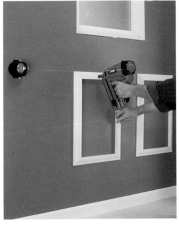

3 SECURE FRAMES TO THE WALL.

Secure the frame to a wall stud if possible. If not, cross-nail as described on page 92. Repeat for the remaining frames. Fill all nail holes (page 96) and apply a finish if desired.

WALL TRIM

Tongue-and-Groove Wainscoting

Wainscoting not only protects walls, but also gives a room a rich finishing touch. Most wainscoting is made up of short vertical boards attached to the lower half of a wall. (Another version, called flat panel wainscoting, is described on pages 136–141). On tongue-and-groove wainscoting (shown here), the edges of the boards are milled with a tongue on one edge and a groove on the

other to join the boards together. Additional trim handles the transitions between the wainscoting and the wall and floor. These include base trim, a cap, and an optional scotia piece that fits under the cap.

To prepare a room for wainscoting, start by removing

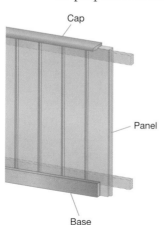

Cap

Panel

Base

any receptacle and switch covers that will be affected. Then pry off the baseboard around the perimeter of the room, as described on pages 68–69. Next, you'll need to create a nailing surface for the boards, as described on the opposite page.

Although the vertical edges of tongue-and-groove wainscoting slip together to create a continuous panel, each board needs to be attached firmly to a wall. The problem: Wall studs are spaced 16" apart on center. This means you'll need to create some type of nailing surface along the full width of each wall. This is most commonly done with backer strips or with backer panels.

BACKER STRIPS. Backer strips are horizontal strips of wood (typically 1×2's) that are attached to the wall studs. To allow the wainscoting to sit as flush as possible to the wall covering, the wall covering is usually cut away.

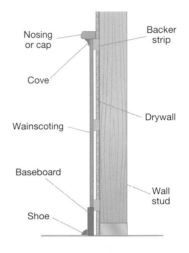

BACKER PANEL. A less messy alternative to cutting grooves in drywall is to simply remove the drywall below the cap and install a continuous plywood backer.

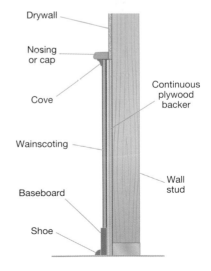

1 **LOCATE THE STUDS.** To install tongue-and-groove wainscoting, begin by locating the wall studs. Use an electronic stud finder to locate these (as described on page 63) and then mark each location with a pencil.

2 **LAY OUT THE BACKER STRIPS.** The next step is to add backer strips or panels. Whichever method you choose, you'll first need to remove drywall. We used three backer strips here, so we marked their locations on the wall. See the sidebar below for a quick way to do this.

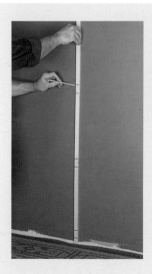

P R O   T I P

Story Stick

Although accurate measurements are good, story sticks are better for laying out trim, as they remove any chance of misreading a measurement. To make a story stick, lay out your trim pattern on a scrap of wood. Now you can use the stick to work your way along a wall to lay out a pattern in seconds—no measurements required.

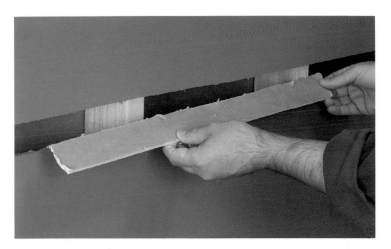

3 **CUT GROOVES IN THE DRYWALL.** Once you've laid out the locations of the backer strips, use a chalk line to snap horizontal lines on the wall. Then cut the drywall and remove the waste. Although you can cut the drywall with a circular saw set to just barely cut through the drywall, it will make quite a mess. A standard drywall saw (page 56) will get the job done more neatly.

4 **INSTALL THE BACKER STRIPS.** Now you can cut the backer strips to length and secure them to the wall studs.

5 INSTALL THE FIRST PIECE. The first piece of wainscoting that you install is the most important, since all other pieces reference off this one. So make sure to use a level when installing it so that it goes in plumb. Secure the piece at each of the backer strip locations.

PRO TIP

Conditioning Wainscoting

Wainscoting made from solid wood expands and contracts as it reacts to humidity. That's why it's important to "condition" your wainscoting prior to installation. This involves "stickering" the wainscoting in the room where it will be installed so that it can acclimate to the room's humidity. Set the pieces on stickers (scraps of ¾"-square dry wood) to let air circulate freely. Most manufacturers suggest allowing the wainscoting to acclimatize for at least two days before installation. If you don't do this, the pieces can shrink or swell once they're installed, creating buckling or gaps.

6 **CONTINUE ADDING BOARDS.** Once the first piece of wainscoting is installed plumb, it's easy going. Just mate the tongue-and-groove edges of additional pieces together and secure them to the backer strips. It's a good idea to stop every three or four pieces and recheck with a level to make sure they're still plumb. If they're not, just shift the next piece slightly to bring it back into plumb. Continue working your way around the room, cutting pieces to fit at the end of each wall as needed.

7 **INSTALL THE BASE TRIM.** Now you can install the base trim. Cut it to length and secure it to the wainscoting and wall studs. Cope the ends as necessary, as described on pages 80–81.

8 INSTALL THE CAP.
The cap or nosing rests on top of the wainscoting and is held in place with nails. Cut pieces to fit, and secure them to the top backer strip or panel. On both inside and outside corners, join the cap with miter joints.

9 INSTALL THE SCOTIA.
Finally, to conceal any gaps between the cap and the wainscoting and to create a smoother transition, cut and install scotia under the cap. Scotia is typically $1/2$" or $3/4$" cove molding. When done, fill all nail holes (page 96) and apply the finish of your choice (pages 98–99).

Quick-Install Wainscoting

In years past, installing wainscoting meant using tongue-and-groove boards or wood panels that were held in place with rails and stiles—something that required advanced woodworking skills. Fortunately, installing wainscoting is much easier today, thanks to new wainscoting products. These new wainscoting "systems" utilize fit-together pieces that are machined on the ends to fit into matching base and cap moldings. This makes installing wainscoting a simple process that can easily be done in a weekend.

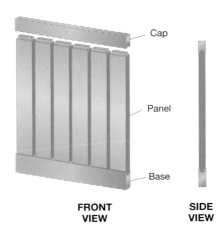

Cap

Panel

Base

FRONT VIEW

SIDE VIEW

1 **INSTALL THE BASE TRIM.** To install quick-install wainscoting, begin by locating and marking the stud locations. Then cut and attach the base trim to the wall at each of the marked stud locations. Check the base trim with a level as you install it to make sure it creates a level foundation for the wainscoting panels. Cut the base as needed when you encounter corners and work your way around the room.

2 **INSTALL THE WAINSCOTING.** Now you can add the wainscoting. With most systems, the pieces slip into a groove in the top edge of the base trim. Since the pieces are tongue-and-grooved together and are also held in place with the cap molding, you don't need to attach each piece to the wall. Use a level to occasionally check the pieces to make sure they're going in plumb; adjust as necessary. Keep adding pieces, attaching them to the wall whenever one is positioned over a wall stud.

3 ADD THE CAP.
When all the wainscoting is in place, add the cap molding. The bottom edge of this should be grooved to fit over the wainscoting. Set the cap in place and check it with a level before securing it to the wall studs with nails. Continue adding cap molding until the wainscoting is complete.

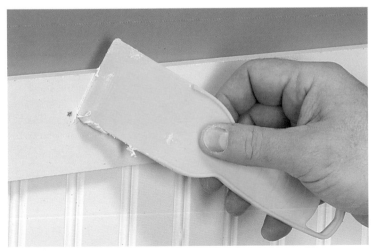

4 FILL HOLES WITH PUTTY.
When all the cap is installed, go around the room and fill in any nail holes with putty; sand flush when dry (page 96). Then apply the finish of your choice, as described on pages 98–99.

Flat-Panel Wainscoting

Of all the different types of wainscoting, flat-panel wainscoting is the most formal looking. That's because it consists of panels set into frames. In the past, the panels fit into grooves cut in the frames, as described on page 140. This took considerable woodworking skill. A simpler version is to attach a frame to a solid back panel and then add molding inside the frames, as illustrated in the bottom drawing.

If our forefathers had had access to engineered panels (like plywood and medium-density fiberboard), they would have used this simpler method. The frames of yesteryear were joined together with stout mortise-and-tenon joints, which also required considerable skill. Fortunately, modern technology—this time in the form of a biscuit jointer (page 48)—makes it easy to join together the frames.

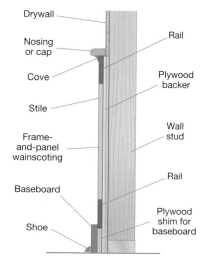

Drywall

Nosing
or cap

Cove

Stile

Frame-
and-panel
wainscoting

Baseboard

Shoe

Rail

Plywood
backer

Wall
stud

Rail

Plywood
shim for
baseboard

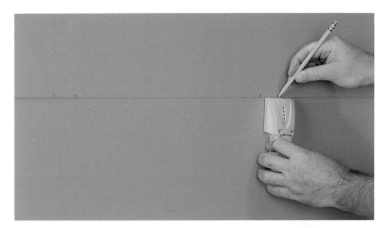

1 LOCATE AND MARK THE STUDS. Wainscoting is usually installed 32" to 36" above the floor. Measure this distance up from the floor and make a mark. Then snap a level chalk line around the perimeter of the room as a reference line for installing the backer panel. Next, use an electronic stud finder to locate and mark the studs as described on page 63.

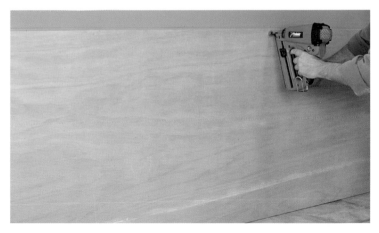

2 INSTALL THE PLYWOOD BACKER. Now you can cut the backer panel (see page 127) to size and secure it to the wall. If you want the wainscoting to fit as flush to the wall covering as possible, remove the drywall behind the panel prior to installing it (see page 127).

3 **LAY OUT THE FRAMES ON THE PANELS.** The next step is to lay out the frames on the panels. You have two options on sizing: You can make all the panels the same size and live with partial panels on one end of each wall, or you can custom-size panels for each wall so that each panel ends up equal in size. (For more on sizing panels, consider using the golden rectangle described on page 124.)

4 **CUT THE FRAME JOINTS.** Once you've laid out the frames and panels on the backer panel, use these dimensions to size your frame pieces. Cut the pieces to size and then join the ends together with biscuit joints. See pages 88–89 for more on working with a biscuit jointer.

5 **GLUE UP THE FRAMES.** With all the biscuit slots cut, apply glue to the slots and biscuits and press the parts of the frame together. Then apply clamps to the frame. Check to make sure the frame is square, and let the glue dry overnight.

6 **ATTACH THE FRAME TO THE PANEL.** Remove the clamps and position the frame on the backer panel. Secure it to the backer panel with glue and nails.

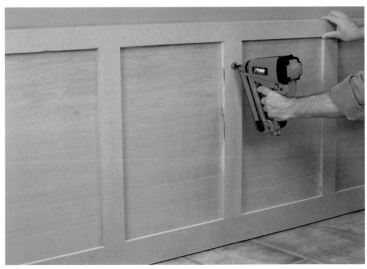

7 **ATTACH THE MOLDING TO THE FRAMES.** Now you can miter molding to fit around the insides of each panel. Secure the molding to the frames with glue and brads.

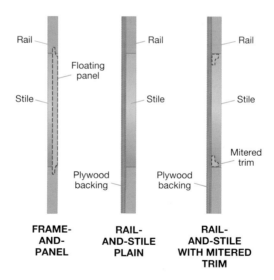

Rail

Floating panel

Stile

Plywood backing

Rail

Stile

Plywood backing

Rail

Stile

Mitered trim

FRAME-AND-PANEL

RAIL-AND-STILE PLAIN

RAIL-AND-STILE WITH MITERED TRIM

PANEL OPTIONS. There are a number of ways you can treat the panels in flat-panel wainscoting. In the past, the panels were made of solid wood and "floated" in grooves cut in the frames; this let the panels expand and contract with the seasons. Modern flat-panel wainscoting with plywood backer can be left plain, or mitered trim can be added, as shown.

8 INSTALL THE NOSING. All that's left is to cut and install the cap or nosing on top of the wainscoting. When all the nosing is in place, go around the room, filling all nail holes as described on pages 96–97. Then apply the finish of your choice, as described on pages 98–99.

All that's left is to cut and install the cap or nosing on top of the wainscoting. When all the nosing is in place, go around the room, filling all nail holes as described on pages 96–97. Then apply the finish of your choice, as described on pages 98–99.

Pre-Made Raised Panels

If you want to add a distinctive touch to your project, consider installing pre-made raised panels onto each of the wainscoting's flat panels. The highly sculpted panels shown here are urethane foam manufactured by Fypon (www.fypon.com). They can be secured easily with polyurethane glue and a couple of brads.

WALL TRIM

Installing a Wall Niche

A pre-made wall niche is about as close as you can get to an "instant" wall makeover. Not only does a wall niche offer distinctive display space, but it can also serve as a handy shelf in an entryway for keys, mail, and other items. Although you'd never know it by looking, the wall niche shown here is made of urethane foam, manufactured by Fypon (www.fypon.com). This particular niche is designed to fit between studs, so installation is a snap.

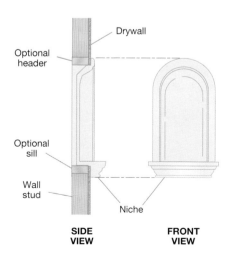

Drywall

Optional header

Optional sill

Wall stud

Niche

SIDE VIEW

FRONT VIEW

1 LOCATE THE STUDS.

Use an electronic stud finder to locate the wall studs where you'll be locating the wall niche as described on page 63. Use a pencil to mark each stud location. Note that if you want to shift the niche to one side or the other of the studs, you need to take precautions to avoid weakening the wall. You'll have to remove the drywall, cut the studs, and install a header and sill plate, as illustrated on page 142.

2 MARK THE CUTOUT ON

THE WALL. The next step is to locate and mark the opening in the wall for the niche. Some manufacturers provide a paper template for this; others don't. If your niche didn't come with a pattern, flip the niche over and measure its back. Then transfer these measurements to the wall or make a paper template, as we did here. Tape the template to the wall and trace around it.

3 CUT OUT THE OPENING. Remove the paper template and cut out the opening. You can do this with a drywall saw, or with a reciprocating saw (see page 57).

4 TEST THE FIT. Now you can test the fit of the niche in the opening. If you cut the opening to match a template provided by the manufacturer, it will probably fit. If you made your own template, it may not fit; simply tweak the opening a bit.

5 **INSERT THE NICHE.** Once the niche fits well in the opening, you can install it. On the niche shown here, there's a large rim around its perimeter. This is an excellent place to apply a bead of polyurethane glue before inserting the niche into the opening. Make sure the rim presses flat against the wall.

6 **SECURE THE NICHE.** Although the polyurethane glue is all you'll need to secure the niche to the wall, it's a good idea to drive a couple of nails through the inside of the niche into the wall studs. Fill these holes (page 96) and apply the finish of your choice (pages 98–99).

Pilasters and Columns

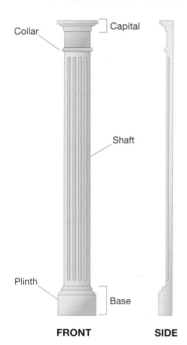

Collar

Capital

Shaft

Plinth

Base

FRONT　　　**SIDE**

Pilasters and columns can add style to any home's interior or exterior. With the advent of easy-to-use urethane foam versions, pilasters and columns can readily be incorporated into a room's design. The pilaster we installed here is manufactured by Fypon (www.fypon.com). It's important to note that decorative foam trim like this does not provide any structural support.

PILASTERS. A pilaster is a shallow column that's attached to a wall or cabinet to create the illusion that it's providing support.

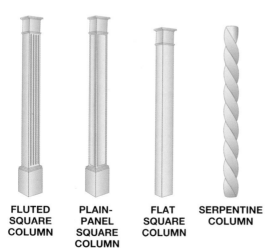

FLUTED SQUARE COLUMN

PLAIN-PANEL SQUARE COLUMN

FLAT SQUARE COLUMN

SERPENTINE COLUMN

COLUMNS. Columns (also called pillars) can be decorative or structural, round or square, tapered or straight. Decorative columns are made of wood or foam. Structural columns are made of solid wood or metal. Plain metal column supports can be wrapped with a decorative column, as described on pages 148–151.

1 LOCATE THE WALL STUDS.

To install a pilaster, begin by locating the wall studs. Use an electronic stud finder to locate these (as described on page 63), and then mark each location with a pencil, as shown in the top left photo.

2 APPLY ADHESIVE.
For a foam pilaster, apply a bead of quality urethane adhesive to its back, as shown in the top right photo. For a wood pilaster, apply a bead of construction adhesive.

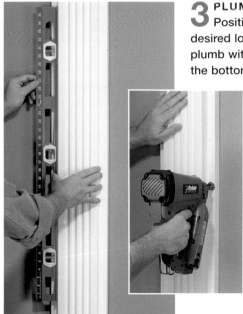

3 PLUMB AND SECURE.
Position the pilaster in the desired location and check for plumb with a level, as shown in the bottom left photo. Then secure it to the wall (bottom right photo). If possible, drive fasteners into a wall stud. If this isn't possible, cross-nail as described on page 92. Fill any nail holes with putty (page 96), and apply the finish of your choice.

Wrapping a Post with a Column

Metal posts and beams that support floors and ceilings are often left exposed—especially in basements—and they're not very attractive. However, they don't have to stay that way. It's a fairly straightforward procedure to wrap an existing pole or beam with a good-looking wood column. A column wrap consists of a plywood sleeve that attaches to the metal pole with a pair of mounting blocks at the top and bottom of the pole. Narrow strips of wood (stiles) are attached to the corners of the box to conceal the plywood edges. To complete the look, add top and bottom rails and some quarter-round.

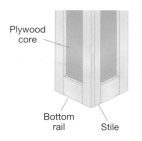

Plywood core

Bottom rail

Stile

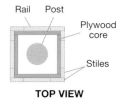

Rail   Post

Plywood core

Stiles

TOP VIEW

1 INSTALL MOUNTING BLOCKS. To wrap a column, start by making a pair of mounting blocks. Size these to fit inside the sleeve that you'll build (our blocks are 4" square), and cut a hole in each to match the diameter of the metal pole. Cut each block in half, and position the split block around the pole. Then secure the blocks to the floor and to the ceiling, taking care to line them up.

2 **BUILD THE PLYWOOD SLEEVE.** Next, cut the plywood sleeve parts to width and length so they're slightly shorter than your floor-to-ceiling measurement. Then fasten together two sides and a back to create a U-shaped sleeve. For maximum strength, we used glue and screws to fasten the parts together.

3 **MOUNT THE SLEEVE.** Now you can slip the sleeve over the mounting blocks and fasten it to the blocks.

4 ADD THE COVER.
Once the sleeve is secured to the mounting blocks, you can add the cover to the sleeve. Here again, we chose screws for their superior holding power.

5 ADD THE STILES.
Now you can cut the stiles to width and length. If you're using ¾"-thick stock (as we did here), you'll cut four stiles ¾" narrower than the other four stiles, as illustrated in the cross-section drawing on page 148. Attach the stiles to the corners of the sleeve with glue and nails.

6 ADD THE RAILS.

Once all the stiles are attached to the sleeve, you can cut the top and bottom rails to fit. For aesthetics, it's best to make the bottom rail about 30 percent wider (taller) than the top rail. Secure these to the sleeve with glue and nails. You may also want to install a middle rail to break up the tall column—this is a matter of personal preference.

7 INSTALL THE TOP TRIM.

Finally, miter some quarter-round or cove molding to wrap around the top and bottom of the column to conceal any gaps between the column and the floor or ceiling. Secure this trim to the column with glue and brads. Fill all nail holes (page 96), and apply the finish of your choice (pages 98–99).

7

Ceiling Trim

QUICK: DESCRIBE THE CEILING above you. You'll probably miss most details, because most of us don't notice ceilings. Usually painted white or off-white, ceilings are easy to ignore. They deserve attention, though, because they can make a big difference in the look of a room. Just by adding simple molding, you can turn that ordinary expanse into a noticeable improvement. In this chapter, we'll show you how to install simple and crown molding, as well as a ceiling medallion.

Simple Molding

One of the best ways to add elegance to a room is to attach simple molding to the ceiling. A basic, narrow molding like the one shown here works especially well in rooms with low ceilings. For higher ceilings, consider installing wider crown molding, as described on pages 158–163. You can attach molding made from natural wood, as shown here. Another popular look is to use paint-grade molding (page 26) and paint it to match the room's accent colors; this makes a striking contrast.

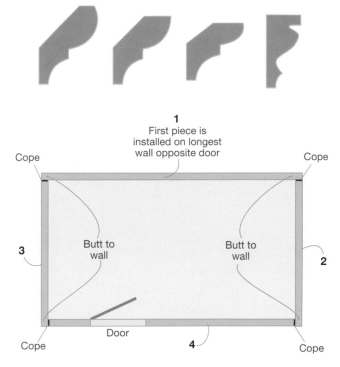

1
First piece is installed on longest wall opposite door

Cope

Cope

Butt to wall

Butt to wall

3

2

Cope

Door

4

Cope

INSTALLATION SEQUENCE. If you're planning on coping the corner joints (see pages 80–81), there's a recommended sequence for installing the molding. This sequence is designed to offer the best possible result.

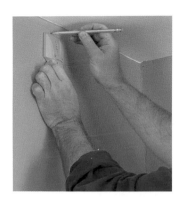

1 **LOCATE AND MARK THE WALL STUDS.** To install simple molding, start by using an electronic stud finder to locate the wall studs, as described on page 63. Then mark these locations with a pencil.

as described on page 63.

Nailing Blocks

A simple way to provide better support to angled molding is to cut a set of mitered nailing blocks and attach these to the wall studs. To make these, measure the flat portion of the molding and miter-cut scraps to create an identical flat on the long side of the block. Attach these at the wall studs. You'll find it's much easier to install and attach the molding to the nailing blocks than it is to attach the molding directly to the wall studs.

CEILING TRIM

2 **BUTT TRIM INTO CORNER.** Start with the longest, most visible wall in the room. Measure and cut a piece of molding to run from corner to corner, butting each square end into a corner.

3 **ATTACH TO NAILING BLOCKS.** Hold the molding in place (see page 161 for tips on holding long pieces of molding in position), and drive fasteners through the molding and into the nailing blocks.

4 **SCARF AS NEEDED.** As you work your way around the room, odds are that you'll need to splice strips of molding together to cross long expanses. When this occurs, miter the ends of the molding in opposite directions to create a nearly invisible "scarf" joint.

5 **COPE AS NEEDED.** Whenever pieces of molding meet at an inside corner, you'll need to cope the joint, as described on pages 80–81. Outside corners are mitered as described on pages 73–75. When all the molding is installed, go back and conceal any nail holes with putty, as described on pages 96–97. Then apply the finish of your choice (pages 98–99).

Crown Molding

Crown molding can dress up any room, adding a graceful touch with its classic profiles. The newer extruded-foam varieties—like the one shown here from Fypon (www.fypon.com)—are much easier to install than conventional wood crown.

There are a few things to keep in mind when using foam trim. First, foam trim should be used for decorative

purposes only; it does not provide any structural support. Second, it's important to check your local building codes to make sure the trim meets local specifications. Third, know that you can paint this trim any color you want: It takes paint well. Some manufacturers now offer embossed molding that looks like wood and can be stained to match other wood in the room. Note that foam molding should never be installed with just fasteners—it's designed to be installed with both fasteners and a bead of high-quality polyurethane adhesive. You don't need a lot: About a ⅛" bead works fine. To install crown molding, you'll first need to locate the studs with an electronic stud finder (page 63) and mark their locations with a pencil.

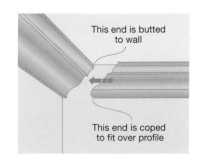

This end is butted to wall

This end is coped to fit over profile

1 INSTALL THE FOUNDATION MOLDING. Because crown rests against a wall at an angle, it has only two small flats that make contact with the wall and ceiling. This makes it difficult to attach. One way to get around this is to first

attach a foundation molding that provides a continuous nailing surface for the crown molding. In addition to providing a nailing surface, a profiled foundation molding also creates a more complex and pleasing profile.

P R O   T I P

Dealing with Uneven Walls

The walls in most homes are not plumb, level, or square. This means that you'll often have to scribe and fine-tune trim to fit (see page 83). This is necessary with

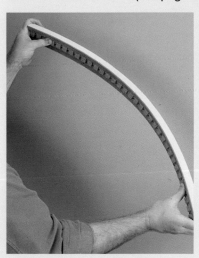

wood molding because it's rigid. Wouldn't it be nice if it was so flexible that you could bend it in and out as needed to hug a wall? You can—if the molding is foam. Foam molding is quite flexible and easily flexes to accommodate imperfections in a wall.

2 **BUTT CROWN INTO A CORNER.** Start by installing crown on the longest wall opposite the doorway leading into the room. Butt the ends of the molding into each corner.

3 **SECURE CROWN TO FOUNDATION MOLDING.** While holding the crown in place (see the sidebar on the opposite page for two easy ways to do this), secure the crown molding to the foundation molding with finish nails. For attaching it by hand, see pages 84–85; to attach it with an air nailer, see pages 86–87.

Holding Crown in Place

Crown molding is a challenge to install, on many levels, the most common being cutting and coping joints. Although attaching crown is fairly straightforward, especially if you use an air nailer, holding it in position to do this is awkward at best. That's because crown molding is often quite heavy and pieces are frequently long. This makes it almost impossible to install by yourself. That is, unless you employ one of the two tricks described below.

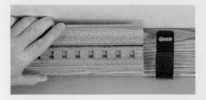

USE CROWN MOLDING CLIPS. Crown molding clips are angled strips of plastic with a keyhole slot at the top. Nails are driven into the wall studs near the ceiling, and the molding clip is slipped over the head of the nail via the keyhole. Then the molding is inserted into the strips. This holds the crown in place for fastening. As you work from one end of the wall, you slide the clip over and lift it off the nail head. Fasten the crown and move to the next clip. It's simple but effective.

USE A "3RD HAND." Another way to hold crown in place is the 3rd Hand, from FastCap (www.fastcap.com). This simple tool is a telescoping pole with pads at both ends. Squeezing the handle near the top lengthens the pole, gently forcing it up against your molding to hold it firmly in place so you can fasten it securely.

4 COPE AND SECURE.
When you reach an inside corner, you'll need to cope the joint, as described on pages 80–81. Once you've got a good fit between the pieces of crown, you can secure the crown to the foundation molding.

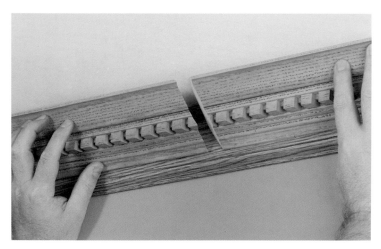

5 SCARF AS NEEDED. As you work around the perimeter of the room, you'll probably need to splice strips of crown together. When this occurs, miter the ends of the molding in opposite directions to create a nearly invisible "scarf" joint.

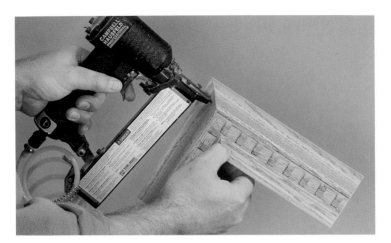

6 **PIN THE OUTSIDE CORNERS.** If you have to wrap crown around an outside corner, it's a good idea to pin together the mitered corners by driving a fastener or two into the top edges of the molding. This will do two things: help pull the miter joint closed, and create a stronger installation.

Burnishing Miters

If you notice small gaps when you mitered together crown molding at an outside corner, you can close the gaps with an old carpenter's trick called "burnishing." All you have to do is press the shank of a screwdriver firmly over the miter joint. This will crush the wood or foam fibers and fill in the gap.

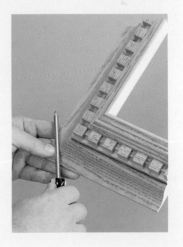

Installing a Ceiling Medallion

A ceiling medallion can add a distinctive touch to a ceiling—with or without a light fixture. Ceiling medallions are also great for concealing an unsightly portion of ceiling left exposed when a larger lighting fixture is replaced with a smaller fixture.

The medallion shown here is manufactured by Fypon (www.fypon.com) and is made from urethane foam, so it's lightweight and easy to install. Urethane foam takes paint readily; so if you'd like to add some color, paint the medallion before mounting it to the ceiling.

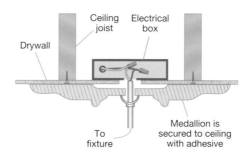

Drywall

Ceiling joist

Electrical box

To fixture

Medallion is secured to ceiling with adhesive

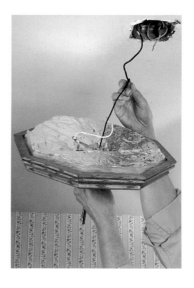

1 REMOVE THE FIXTURE. Start by turning off the power to the fixture. Then remove the diffuser and the lightbulb(s). Next, unscrew the retaining nut that holds the decorative cover plate onto the electrical box. If the old fixture doesn't come off easily, run the blade of a putty knife around the edges of the cover plate to free it from old caulk or paint. Then unscrew the wire nuts, separate the wires, and set the old fixture aside.

2 APPLY ADHESIVE.

Because it's so light and offers such a large gluing surface, a urethane medallion can be secured to the ceiling just by applying a bead of high-quality urethane adhesive to its back.

3 PRESS IN PLACE.

To install the medallion, simply press it in place over the electrical box. Apply a couple of strips of tape to keep it in place until the adhesive sets up.

4 INSTALL THE FIXTURE.

Now you can attach the fixture wires to the circuit wires with wire nuts, and attach the fixture to the mounting plate. Then screw in the appropriate bulbs and attach the diffuser. The diffuser is usually held in place with a decorative cap or retaining nut. Tighten this to friction-tight and no more—overtightening can crack the diffuser.

8

Windows and Doors

WANT TO OPEN UP a completely different look for a room, or spruce up that first impression made by your entryway? Consider a new window or door to make a real change in more than just appearance. These enhancements can let the sunshine in, change a traffic pattern, boost your sense of privacy and security, save on energy costs, and ultimately add to the value of your home. On a more modest scale, if appearance alone is the issue, you can spruce up existing windows and doors with new trim.

Removing a Window

If you're planning on replacing a window, you'll first have to remove the old one.

1 **REMOVE THE INTERIOR TRIM.** The first step to removing a window is to remove the interior trim. Before you break out a prybar or putty knife, take the time to slice through any paint between the trim and the wall with a utility knife. In many cases, especially older homes where the walls and trim have multiple layers of paint, paint can form a strong bond between the trim and the wall. If you don't cut this paint "joint," you'll likely damage the wall covering when you remove the trim.

2 **REMOVE THE EXTERIOR TRIM.** The next step is to remove any exterior trim. A prybar will make quick work of this, as described on page 68. If you're planning on reusing the trim, protect it from damage from the prybar by slipping a stiff putty knife behind the trim. Then remove the nails as described on pages 68–69.

3 RELEASE THE WINDOW. Once you've removed both the interior and exterior trim, you'll have access to the framing members. If the window is screwed in place (as shown in the top right photo), simply remove the screws. A quick way to release a window is to cut through its nails with a reciprocating saw, fitted with a demolition blade (top left photo). A couple of things here: First, make sure you have plenty of clearance for the blade—too tight an opening will result in kickback and bent blades. Second, keep the guide of the saw pressed firmly against the wall as you cut, and keep a firm grip when you encounter a nail.

4 REMOVE THE WINDOW. At this point the window should be free and ready to remove. If it isn't, stop and find out what's holding it in place. Quite often there's a hidden nail or screw that you missed. For larger windows, have a helper on hand as you pull it out of the rough opening.

Installing a Window

The most common type of replacement window is a double-hung window, illustrated in the drawing at right. On a double-hung window, both the upper and lower sashes slide up and down. Regardless of the type of window, the basic steps for installation are similar: You insert the window in the rough opening, shim it so it's level and plumb, and secure it (see page 172 for more on mounting options).

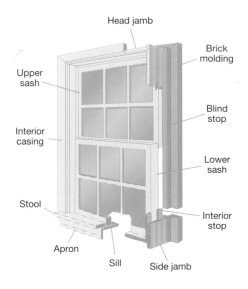

Head jamb

Brick molding

Upper sash

Blind stop

Interior casing

Lower sash

Stool

Interior stop

Apron

Sill

Side jamb

Window Wrap

Before you install a window in a rough opening, it's a good idea to create a weatherproof seal around the opening. There are a number of self-adhesive and staple-on membranes or "wraps" designed specifically for this. They're made to slip under the siding and wrap around the framing of the rough opening, as shown.

1 **TEST THE FIT.** If you measured correctly and have ordered the correct size replacement window, the new window should slide in easily, as shown in the top photo. You should have about ½" to 1" combined clearance between the sides of the new window and the jamb. This space allows you to slip in shims and level the window; see below.

2 **SHIM AND LEVEL THE WINDOW.** In order for the new window to operate without binding, it's important that it be installed level and plumb. Check for this with a level on the sides, top, and bottom. Insert pairs of shims as needed in the gaps between the sides of the new window and the jambs. Slide the shims in and out until the window is level and plumb.

WINDOWS AND DOORS

3 SECURE THE WINDOW.

With the window level and plumb, you can now secure it to the jamb with casing nails or screws. To keep from bowing the sides of the window, make sure to drive in nails or screws only where there are shims. Cut off any protruding shims with a sharp utility knife.

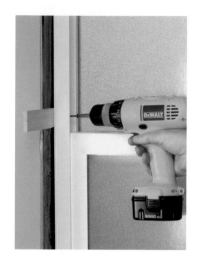

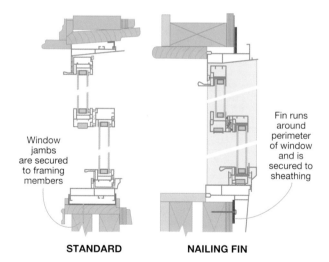

Window jambs are secured to framing members

Fin runs around perimeter of window and is secured to sheathing

STANDARD   **NAILING FIN**

MOUNTING OPTIONS. Windows are mounted in one of two ways: through the jambs or via nailing fins. Traditional windows were attached by driving screws or nails through the jambs and shims into the rough opening framing. Most new construction windows come with perimeter flanges that are nailed to the exterior sheathing, as shown.

4 SEAL THE WINDOW. Since windows are installed in oversized rough openings, there will be gaps between the window jambs and the framing. Although these gaps will be covered by trim, the gaps can and will allow air to flow in and out of the house. To prevent warm air from escaping in the winter and cool air in the summer, you need to fill these gaps. This can be done with fiberglass insulation, foam rod (as shown here), or low-expanding foam, as described on page 34.

5 ADD THE TRIM. Finally, you can add the interior and exterior trim. Measure and cut the trim pieces to length and attach them to the jamb and framing members. Fill all nail holes with putty, caulk around the trim to seal any gaps, and apply the finish of your choice.

WINDOWS AND DOORS

Exterior Window Treatments

Go from boring to brilliant. That's what you can do to a window by just adding some trim, as shown in the photos below. What's really nice about the exterior trim shown here is that it's so easy to install—and virtually impervious to weather. How is this possible? The trim is urethane foam, manufactured by Fypon (www.fypon.com). It cuts easily and goes up with a bead of polyurethane adhesive and a couple of fasteners. Foam trim is available in a huge variety of shapes, sizes, and styles: from simple molding to arched or decorative window panels.

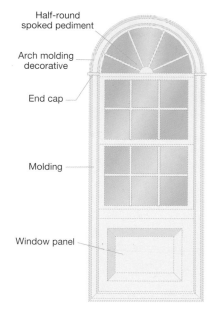

Half-round spoked pediment

Arch molding decorative

End cap

Molding

Window panel

1 **ATTACH THE SIDE TRIM.** To install exterior window trim, first remove any existing trim (if applicable). Then follow the manufacturer's directions to measure and cut the side trim to size. Apply a bead of polyurethane adhesive and secure each side with fasteners.

2 **INSTALL THE APRON.** Next, cut the apron to length and secure it with polyurethane adhesive and fasteners, as shown in the lower left photo.

3 **ATTACH THE HEADER AND KEYSTONE.** Finally, cut the header to size and attach it and the keystone (if applicable) with polyurethane adhesive and fasteners, as shown in the right photos.

Trimming a Wall Opening

Pass-through openings in walls are often left plain. In many cases the opening is finished with drywall only, with no attempt at trimming the opening to match the surrounding room interiors. In years past, this required considerable woodworking skill. But with the pre-made parts (like plinth blocks, fluted trim, and rosettes) that are now commonly available at most home centers and lumberyards, trimming an opening is a fun and easy project. To trim a wall opening, you'll first need to install a jamb, as described below.

1 **INSTALL A JAMB.** Besides providing a finished look for the wall opening, a jamb also provides a nailing surface for the wall trim that you'll add later. You can either purchase a pre-made jamb or make one yourself (as described on the opposite page). In either case, you'll likely need to trim the jamb pieces to length. Fit the top jamb between the side jambs and fasten them together with nails. Then slide the jamb into the opening and secure it to the framing members.

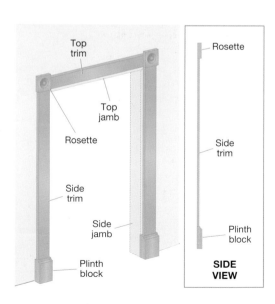

Top trim

Top jamb

Rosette

Rosette

Side trim

Side jamb

Plinth block

Rosette

Side trim

Plinth block

SIDE VIEW

Making a Jamb

The jamb for a door or wall opening consists of two side jambs and a top jamb. The top inside faces of the side jambs are rabbeted to accept the top. The rabbets make aligning the pieces easier and also help to create a stronger joint. You can buy pre-made jambs, but it's also easy to make your own. Making your own may be the only choice for homes where the walls aren't standard thickness.

1 CUT THE RABBETS. To make a jamb, first measure the width of the wall. Then measure the height and width of the opening and cut stock to match the width of the wall and 2" to 3" longer than the opening measurements. Next, cut  or rout rabbets on the top inside faces of the side jambs with a router fitted with a rabbeting bit. You can also cut these with a dado blade in a table saw, or by hand.

2 ASSEMBLE THE JAMB. Cut the side jambs to length and then cut the top jamb to fit between the rabbets in the side jambs. Apply glue to the rabbets and insert the top jamb between the side jambs; fasten them together with nails.

2 INSTALL THE PLINTH BLOCKS.

With the jamb in place, the next step is to install plinth blocks at the base of the opening. These can either be installed flush with the jamb or offset ⅛" or so—it's really a matter of personal preference.

3 ATTACH THE ROSETTES.

Now you can install the rosettes at the top corners of the jamb. Make sure that these go in level and plumb with the plinth blocks. A plumb bob is a great way to check alignment. Alternatively, if you offset the plinth block, you can mark the same offset or reveal on top of the jamb.

4 **INSTALL THE SIDE TRIM.** With the plinth blocks and rosettes in place, you can install the side trim. Measure from plinth block to rosette, and cut pieces of trim to fit. Install these so that they are centered on the width of the plinth blocks and rosettes. Fasten the trim to the jamb with nails.

5 **ATTACH THE TOP TRIM.** All that's left is to cut and install the top trim piece. Measure from rosette to rosette and cut a piece to fit. Center it on the rosettes and attach it with nails. Repeat this process for the other side of the opening. Then fill all nail holes (page 96) and apply the finish of your choice (pages 98–99).

Removing a Door

If a new door is in your future, odds are that you'll be replacing a door, not installing one where there was no door before—that rather ambitious project is generally best left to a professional. Replacing a door begins with removing the old one.

1 **REMOVE HINGE PINS.** To remove a door, start by removing its hinge pins, as shown in the upper left photo.

2 **REMOVE THE DOOR.** Now you can detach the door. Swing the door open slightly to get a good grip, and lift the door up slightly and out. Solid-core doors are surprisingly heavy, so have a helper on hand to detach and move the door.

3 **REMOVE THE INTERIOR TRIM.** The next step is to pry off the interior trim, as shown in the bottom photo and described on page 68.

4 REMOVE THE EXTERIOR TRIM. Now remove the exterior trim. In most cases, the exterior trim is brick mold and is quite stout. This means that you'll usually have to resort to a prybar to release it from the jamb.

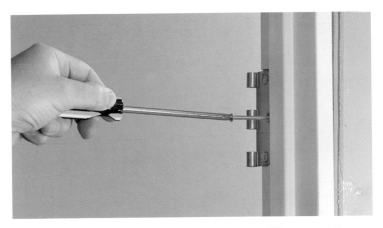

5 REMOVE THE LONG HINGE SCREWS. When most doors are installed, one screw on each hinge jamb is removed (usually the center one) and replaced with a longer screw that will pass through the jamb and into the rough opening framing to firmly support the door. Instead of trying to cut through these screws in the next step, it's easier to simply remove them. If in doubt as to which hinge screws are the long ones, just remove them all.

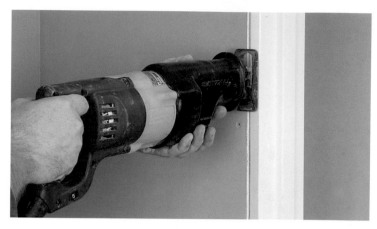

6 **RELEASE THE DOOR.** To release a door jamb, cut through the nails that secure it to the framing members. The best tool for this job is a reciprocating saw, fitted with a demolition blade. Since these saws are powerful and can buck when they come in contact with a fastener, it's important to press the cutting guide of the saw firmly against the wall as you cut. An alternative to cutting the nails is to pry them out. You'll need a cat's paw for this (page 56) and a lot of patience.

7 **PULL OUT THE JAMB.** At this point the jamb should be free of the framing members and can be pulled out of the opening. Grip one of the side jambs firmly and give it a tug to pull it out. After the jamb is removed, you can easily break it down for reuse or disposal.

Installing an Exterior Door

The impact of a new door on the look and feel of a room can be dramatic. For example, if you remove an exterior solid-core door with no windows and replace it with a new door that has windows (as shown here), light will enter the room, providing an entirely different ambiance.

Before you install your new door, you'll need to first remove the old one; see pages 180–182 for step-by-step instructions on how to do this. On new exterior doors, the exterior brick mold may be pre-installed, or you may have to attach it to the doorjamb prior to installation (see below). Brick mold is much thicker than interior door casing because it was originally used to provide an edge for bricks to butt up against on homes with brick exteriors. Although it's used for all types of exteriors now, the name has stuck—along with the unnecessary thickness.

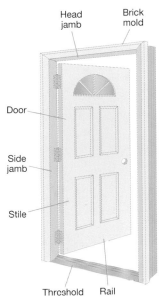

Head jamb
Brick mold
Door
Side jamb
Stile
Threshold
Rail

1 INSTALL THE EXTERIOR TRIM. Brick mold is available pre-cut and pre-primed. All you need do is mark the reveal (page 64) and nail it to the jamb.

2 **TEST THE FIT.** With the brick mold in place, lift the door up and into the rough opening to make sure it fits well. Exterior doors tend to be rather heavy, so have a helper on hand to assist you as you lift and position the door. Once the door is in place, check to make sure you have sufficient clearance between both side jambs and the jack studs for the shims you'll use later to plumb and level the door.

Sealing Thresholds

To create a water-tight seal between a door's threshold and the underlying subfloor, pros apply 100% silicone caulk to the subfloor before installing the door. They know that it's a whole lot easier to wipe off excess caulk than to remove a door after it's been installed when leaks are discovered, and apply more caulk.

3 **LEVEL AND PLUMB THE DOOR.** Start by inserting shims behind each of the three hinges, behind the opening for the plunger for the door lockset, and at the top and bottom of the latch-side jamb. Insert the shims in pairs of opposing wedges, and adjust them in and out until they solidly fill the gap between the jamb and the framing members. Hold a 4-foot level up against one of the side jambs and check it for plumb. Adjust the position of the shims until the door is plumb. Then move to the other side jamb and repeat the process.

4 **SECURE THE DOOR.** Now you can secure the door with 2½"- to 3"-long galvanized casing nails. Make sure to drive the nail through the jamb only where the shims are located. You want to drive the nail though the jamb and the shims and into the framing members. This way the jamb will be fully supported.

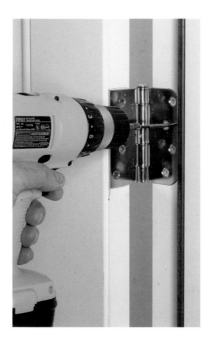

5 INSTALL THE LONG HINGE SCREWS.

Most pre-hung doors come with three long hinge screws that are designed to fasten the jambs to the framing. The door hinges may or may not have an empty slot waiting for these. If not, you'll need to remove one screw from each jamb hinge and replace it with a longer screw.

6 SEAL THE DOOR.

Since doors are installed in oversized rough openings, there will be gaps between the jambs and framing. Although these gaps will be covered by trim, the gaps can and will let air flow in and out of the house. To prevent warm air from escaping in the winter and cool air in the summer, you will need to fill these gaps. This can be done with fiberglass insulation (as shown here) or low-expanding foam (see page 34).

7 SECURE THE EXTERIOR TRIM.

The next step is to secure the brick mold to the framing. Drive nails through the face of the brick mold and into the framing members about every 12" or so. Then seal the exterior trim, as described on page 93.

8 INSTALL THE INTERIOR TRIM.

Now you can finish off the inside of the door. Start by cutting off any protruding shims with a sharp utility knife. Then cut and attach interior casing. Finally, install the lockset of your choice and test the operation of the door. If all went well, it should open and close smoothly without binding and create a solid seal against the elements.

WINDOWS AND DOORS

Exterior Door Treatments

Almost every exterior door can benefit from added trim. Look at the amazing difference between the before and after photos shown below. Although it looks like the trim was handmade by a skilled woodworker, it's made of foam: It's super easy to install and virtually impervious to weather. The trim is manufactured by Fypon (www.fypon.com). It cuts easily and goes up with a bead of polyurethane adhesive and a couple of fasteners. Foam trim is available in a huge variety of shapes, sizes, and styles—everything from simple molding to arched or decorative side panels, as illustrated below.

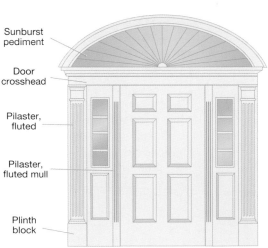

Sunburst pediment

Door crosshead

Pilaster, fluted

Pilaster, fluted mull

Plinth block

1 **INSTALL THE SIDE TRIM.** The foam trim shown here is designed to butt up against existing brick mold. To install it, start by measuring and cutting the side pilasters to length. The ones shown here run from the top of the brick mold to its bottom. Apply a bead of polyurethane adhesive to the back of the pilaster, align it with the top edge of the brick mold, and fasten it in place with galvanized nails.

2 **INSTALL THE BASE.** The pilasters shown are two-piece units consisting of a pilaster and a base. The base fits over the cut end (bottom) of the pilaster and provides a finished look. Install a base flush with the bottom of each pilaster by applying a bead of polyurethane adhesive and driving in fasteners from the side.

3 INSTALL THE HEADER. The next step is to install the header. The foam header we installed comes long and has to be cut to length— basically you have to remove a center section, leaving the desired-length header. To prevent water from becoming trapped behind the header, the manufacturer suggests that you install metal flashing (called drip edge) under the

siding above the header. The flashing fits over the header and creates a watertight seal (see the manufacturer's installation instructions for more on this).

4 ADD THE OPTIONAL KEYSTONE. To conceal the splice in the header, you can install a keystone, as described on page 191.

Adding a Keystone

■■ In masonry, a keystone is the central wedge-shaped
■■ stone of an arch that locks its parts together. In
house trim, a keystone can be purely decorative and can be
installed on both arched and straight headers. In addition to
being decorative, a keystone can be used to conceal an
underlying joint, like the joint between the two foam header
pieces shown here.

**1 INSTALL ONE-HALF OF
THE HEADER.** Follow the
manufacturer's instructions for cutting the header to length. Then
apply a bead of polyurethane
adhesive, position the header so it
extends the desired distance past
the top of the pilaster, and drive in
fasteners.

**2 INSTALL REMAINING HALF
OF THE HEADER.** Next,
apply a bead of polyurethane adhesive to the remaining header half.
Position the piece so it extends the
desired distance past the top of
the pilaster, and drive in fasteners.

3 ATTACH THE KEYSTONE.
Apply a bead of polyurethane
adhesive to the back of the keystone, and center it from side to
side on the header. Then drive in
fasteners. Fill all nail holes (page
96) and apply the exterior finish of
your choice.

WINDOWS AND DOORS

Installing an Interior Door

Installing a pre-hung interior door is similar to installing an exterior door (pages 183–187). The big difference is that you don't have to worry about protecting the door and framing members from the elements. Note that an interior door doesn't have a threshold like the exterior door shown on page 183 (top drawing). Interior doors are easier to install, especially if you buy a pre-hung door. Before you install your new door, you'll need to first remove the old one; see pages 180–182 for instructions on how to do this.

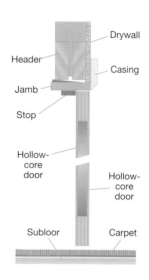

Drywall

Header

Casing

Jamb

Stop

Hollow-core door

Hollow-core door

Subloor

Carpet

1 SHIM AND LEVEL THE DOOR. To install an interior door, start by checking the fit to make sure that you have clearance between the jambs and the studs for shims. Insert shims behind each of the three hinges, behind the opening for the plunger for the door lockset, and at the top and bottom of the latch-side jamb. Hold a level up against one of the jambs and check it for plumb. Adjust the shims until the jamb is plumb. Then move to the other jamb and repeat.

2 **SECURE THE DOOR.** To secure the door, drive nails through the jamb and the shims into the framing members. Your door should have come with three long hinge screws used to fasten the hinge-side jamb firmly into the framing members. The door hinges may or may not have an empty slot waiting for these. If not, remove one screw from each jamb hinge and replace it with a longer screw.

3 **ADD THE TRIM.** Measure and cut door casing to fit, and secure it with nails. Then fill all nail holes (page 96) and install the lockset.

9

Troubleshooting and Repair

ONE OF THE NICE THINGS ABOUT TRIM is that once it's installed properly, it rarely poses any problems. The key word here is *properly.* Improperly installed trim can create problems, and we'll cover how to handle those in this chapter. We'll also look at some common installation problems and their solutions.

Working with Jambs

Improperly installed jambs can create several headaches. If they're not installed flush with the wall covering on both sides of the wall, the trim that's installed to conceal the gap between the jambs and the wall covering won't go on correctly. If the jambs are the correct width to match the walls but were simply installed wrong, your best bet is this: Free the jamb from the wall by cutting through the fasteners, as described on page 182, and re-install the jamb so it's flush with both walls. Jambs that are under- or oversized, as illustrated in the drawing below, must be treated differently.

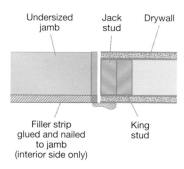

Undersized jamb

Jack stud

Drywall

Filler strip glued and nailed to jamb (interior side only)

King stud

UNDERSIZED JAMB

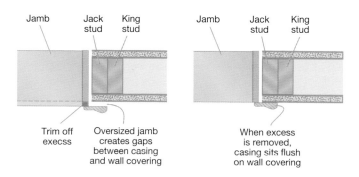

Jamb

Jack stud

King stud

Trim off execss

Oversized jamb creates gaps between casing and wall covering

Jamb

Jack stud

King stud

When excess is removed, casing sits flush on wall covering

OVERSIZED JAMB

UNDERSIZED JAMBS. There are two ways to deal with undersized jambs: Replace them with the correct width jambs, or add filler strips. By far the easier solution is to add filler strips. To do this, measure the difference between the jamb and the wall. Then cut strips of wood to match this and glue and nail them to the jamb.

TRIM OVERSIZED JAMBS. If you can't reinstall or replace an oversized jamb, you'll have to trim it. A block plane or jack plane (page 46) is your best bet here. To keep from damaging the wall covering as you plane, consider running a piece of masking tape along the edge of the jamb. If you tear into this with the plane, you'll know you're close and can stop planing. If you like, switch to a sanding block to finish the job.

OR SHIM THE TRIM. If the trim has already been installed and you don't want to remove it and trim the jamb, you can cut thin shims to fit between the trim and the wall covering, as illustrated in the drawing on page 196. Hold the shims in place with glue and brads.

Troubleshooting Windows

There are three problems commonly associated with windows: The window binds, the sash is loose, or the window is drafty.

Window binds

Windows that bind are more of a nuisance than anything. In most cases, the bind is caused by the stops that hold the sash in place.

APPLY PARAFFIN TO THE STOPS. If a sash binds a little but can still be raised and lowered, try rubbing some paraffin or paste wax onto the stops. You want to apply the wax to the face of the stop that rubs up against the sash, as shown.

SHAVE THE STOPS. If paraffin doesn't do the trick, you'll have to shave the stops a bit to create more clearance between the sash and the stops. A block plane will usually get the job done, or you can use a block of wood wrapped with sandpaper. Alternatively, you can pry off the stop and move it away from the sash.

Sash is loose

A sash that is loose and/or rattles when the wind blows is both annoying and costly. A loose sash does not create a seal against the elements. Not only will you lose warm air in the winter and cool air in the summer, but a loose sash can also let moisture penetrate the wall, resulting in water damage.

SHIM THE STOPS. You can fix a loose sash by shimming the stop with a strip of wood or by applying V-type weatherstripping, as described below.

Window is drafty

When the wind howls through a window, it's a sure sign that either the window needs weatherstripping or the existing weatherstripping needs replacing.

REPLACE OR INSTALL WEATHERSTRIPPING. There are two types of weatherstripping that work well in windows: V-strips of cushion vinyl, and spring-metal strips. Both install between the edge of the sash and the jamb, as shown. Not only will these create an excellent seal, but both also can remove any play or slop in the sash.

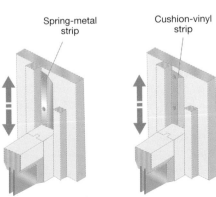

Spring-metal strip

Cushion-vinyl strip

DOUBLE- OR SINGLE-HUNG WINDOWS

Troubleshooting Doors

Since a door has fewer moving parts than a window, it's easier to troubleshoot. The two most common door problems you'll encounter are binding and drafts.

Door binds

If the entire edge of a door rubs up against the jamb, you'll need to either plane the full length of the door (see page 201) or deepen all of the hinge mortises. If the latch edge of the door rubs just at the top or bottom, deepen the corresponding hinge mortise. If the top or bottom edge of the door binds, you can solve the problem either by shimming the appropriate hinge (see below) or by planing off the offending area.

SHIM THE HINGES. Shimming a hinge basically tilts the door just a bit. To do this, remove the hinge screws and insert a piece of stiff paper or cardboard behind the hinge, as shown. Then drive the screws back in and check the operation.

Q U I C K   F I X

Stripped Screws

When working with hinge screws, you may encounter a hole that has been stripped. Here's an easy way to fix this: Dip the end of one or two wooden toothpicks into glue and insert them into the stripped hole. Snap the end off so it's flush and then drive in the screw. The toothpick will give the threads of the hinge screw something to bite into as it's driven home.

SHAVE THE DOOR.

If you need to shave the entire edge of a door, we suggest a block plane or jack plane (page 46). If you need to take off just a little bit, a block wrapped with sandpaper will do. Whichever you choose, stop frequently and test the door.

Door is drafty

If you can see daylight around the perimeter of a door or feel a draft, you need to either replace the existing weatherstripping or install new weatherstripping. The weatherstripping you'll use to seal a door is the same as that used for a window (see page 199). There are a couple of other products that may work well for you, including self-adhesive foam weatherstripping, as illustrated in the drawings below. The bottom of a door requires special treatment in the form of a door sweep, shoe, or metal threshold. A door sweep attaches to the bottom face of the door, and its rubber gasket drags along the threshold to create a seal. A door shoe fits on the bottom of the door itself, its flexible gasket pressing tightly against a threshold. A metal threshold with a rubber gasket is designed to press tightly against the bottom of the door to stop drafts.

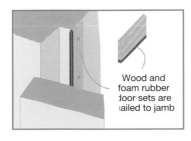

Wood and foam rubber door sets are nailed to jamb

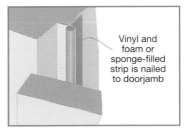

Vinyl and foam or sponge-filled strip is nailed to doorjamb

Troubleshooting Trim

Problems with trim include gaps, cracks, and bowing. Fortunately, the remedies are simple.

Gaps

In a perfect world, casing will lie flat against both the wall and the jamb. In reality, there are often gaps. Two ways to treat these are with shims and with caulk.

SHIM THE TRIM. Gaps between casing and jambs can be eliminated by shimming the back of the casing. In effect, you're tilting the trim forward to close the gap. The downside to this is that you create a gap on the back of the casing—but this can be easily filled with caulk (see below).

FILL THE GAPS. Small gaps can be filled with caulk. Since you'll probably be painting the trim, use a paintable caulk. You'll also want the caulk to flex so it can move with the trim over time as it reacts to seasonal humidity—so go with a paintable latex caulk with silicone.

Cracks

With seasonal changes in humidity, solid-wood trim will expand and contract. Since the trim is securely fastened to the wall, ceiling, and/or jamb, it's not free to move. So, it'll crack. You can fill cracks with putty

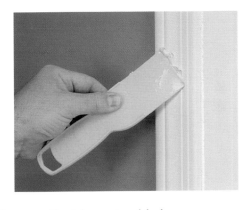

(page 96), but you'll be better off with a paintable latex caulk with silicone. This way, if the wood continues to move, the caulk can flex along with it.

Bowing

Warping or bowing in solid-wood trim is another reaction to seasonal changes in humidity. Here again, since the trim is fastened in place, it can't move freely. Often, then, the trim bows out.

The only thing you can do here (outside of removing and replacing the trim) is to refasten the bowed section to the wall or jamb. When you have the trim as flat as possible, fill any gaps with paintable latex caulk with silicone.

Finish Problems

One of the most common problems with trim concerns the trim's finish. This includes nail pop, stains, and raised grain. All of these are easy to fix.

NAIL POP. All wood moves as it reacts to seasonal changes in humidity. When a nail is driven into wood and the wood expands and contracts, this movement squeezes the nail and tries to force it out of the wood. The result is known as nail pop. If the nail was set and covered with putty, the movement forces the putty right out of the hole. The solution? Scrape away the putty, reset the nail, and apply fresh putty (page 96).

STAINS. Stains are also a common finish problem that can be caused by several things. On softwood trim (like pine), defects such as knots can "bleed" through a finish. The remedy for this is to remove the old finish (see the sidebar on the opposite page) and seal the unsightly area with a coat of shellac. Let the shellac dry, and then repaint. Stains can also be caused by fasteners exposed to moisture. Non-galvanized nails can rust and bleed through a finish. So make sure you always use galvanized fasteners on exterior window and door trim.

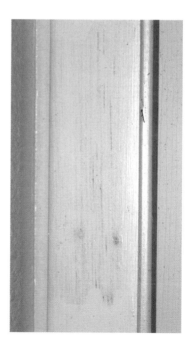

RAISED GRAIN. Raised grain or ridges can occur on solid-wood trim that's not fully dry when installed. Over time the trim will dry completely; but because its surface is sealed with paint, it can't dry uniformly, and ridges can appear. The solution is to remove the finish, sand the trim smooth, and repaint.

Repairing a Finish

 How you repair a finish will depend on the type of finish: clear topcoat or paint.

CLEAR TOPCOAT. To repair a clear topcoat, sand away the finish with a piece of 150-grit silicon-carbide sandpaper. Make sure to feather the edges where the topcoat transitions to bare wood. Then apply a fresh finish. A wipe-on polyurethane is best for this since it doesn't leave brush marks.

PAINT. Blemishes on paint are best removed with aluminum oxide open-coat sandpaper. Be sure to feather the

transition from paint to bare wood. Apply a primer to the bare wood and let it dry. Then brush on a fresh coat of paint.

TROUBLESHOOTING AND REPAIR

Index

Metric Equivalency Chart

Inches to millimeters and centimeters

INCHES	MM	CM	INCHES	CM	INCHES	CM
1/8	3	0.3	9	22.9	30	76.2
1/4	6	0.6	10	25.4	31	78.7
3/8	10	1.0	11	27.9	32	81.3
1/2	13	1.3	12	30.5	33	83.8
5/8	16	1.6	13	33.0	34	86.4
3/4	19	1.9	14	35.6	35	88.9
7/8	22	2.2	15	38.1	36	91.4
1	25	2.5	16	40.6	37	94.0
1 1/4	32	3.2	17	43.2	38	96.5
1 1/2	38	3.8	18	45.7	39	99.1
1 3/4	44	4.4	19	48.3	40	101.6
2	51	5.1	20	50.8	41	104.1
2 1/2	64	6.4	21	53.3	42	106.7
3	76	7.6	22	55.9	43	109.2
3 1/2	89	8.9	23	58.4	44	111.8
4	102	10.2	24	61.0	45	114.3
4 1/2	114	11.4	25	63.5	46	116.8
5	127	12.7	26	66.0	47	119.4
6	152	15.2	27	68.6	48	121.9
7	178	17.8	28	71.1	49	124.5
8	203	20.3	29	73.7	50	127.0

mm = millimeters   cm = centimeters

PHOTO CREDITS